Bioenergetics and Energy Metabolism of Crustaceans

By: L.V.K.S. Bhaskar

Copyright © 2016 by L.V.K.S. Bhaskar

All rights reserved. No part of this publication may be reproduced, distributed, or transmitted in any form or by any means, including photocopying, recording, or other electronic or mechanical methods, without the prior written permission of the publisher, except in the case of brief quotations embodied in critical reviews and certain other non-commercial uses permitted by copyright law.

Contents

Preface	pg i
Introduction	pg 1
Materials and methods	pg 30
Chapter One - Allometry	pg 59
Chapter Two - Characterization of moult cycle	pg 69
Chapter Three - Bioenergetics	pg 83
Chapter Four - Energy metabolism	pg 107
Summary and conclusions	pg 141
References	pg 147

PREFACE

Decapods are of immense food value. They have become an important source for the foreign exchange earners. Prawn fishery has improved remarkably in India with the encouragement of the Government. Though much work has been carried out on crustaceans regarding different aspects of biology and physiology, relatively fewer attempts have been made to understand bioenergetics and related metabolic changes upon eyestalk ablation in Natatians, such as *Macrobrachium malcolmsonii* (H. Milne Edwards). Biology of commercially important animals highlights its position and suitability for culture practices. Of late predicting growth of commercially important crustaceans inhabiting diverse ecosystems has attracted the attention of several investigators (Limburg et al. 1998). For commercial exploitation of a species its bionomics are to be understood in detail and attempts should be made to culture and breed artificially.

The life of a typical decapod crustacean alternates between an intermoult period during which it feeds actively and a short moulting period during which it sheds its old exoskeleton (ecdysis) and increases in size. Between intermoult and moult, a stage intervenes, when the changes necessary for moulting process are accomplished (Premoult). Since, moult cycle is a primary concern in most of the studies on crustacean physiology it is necessary to establish a coherent and explicit scheme of identifying various stages of the moult sequence. Maintenance of intermoult, various metabolic biochemical aspects connected with intermoult, moulting process *per se* and aspects of reproduction are under neuroendocrine control in crustacea.

Bioenergetics is the study of balance between energy supply (through feed) and energy expenditure (Sibly and Calow, 1986) and requires an examination of physiological processes through which energy is transformed in living organisms. Studies on the flow and allocation of energy in biological systems often have, as their ultimate goal, the construction of a budget describing the partitioning of energy within an individual, population (or) community. Such budgets are then used to make inference about physiology or ecology in particular explaining why an organism or population does what it is observed to do assuming that energy allocation is the overriding concern (Sibly and Calow, 1986). Bioenergetic modelling is an alternative method for estimating feeding rates in a natural population. These models quantify the relationship between feeding rates and growth relative to temperature, body size and activity. Bioenergetic models have been applied to address ecological questions in a wide variety of taxa including insects (Wightman, 1981), fish (Kitchell et al. 1977), amphibians (Jaeger and Barnard, 1981), reptiles (Congdon and Tinkle, 1982) and birds (Hainsworth et al. 1981).

Influence of eyestalk hormones as on moulting, growth, reproduction, development, pigmentation, ionic regulation and metabolism of different crustaceans has been studied in great detail (Huberman, 2000). Though enough information is available on the effects of eyestalk ablation in crabs and to a lesser extent in shrimps and prawns, literature pertaining to bioenergetics and metabolism of carbohydrate, protein and lipid upon eyestalk ablation appears rather insufficient and this is particularly so in the case of *Macrobrachium* species. Therefore the present work, primarily concerns with the estimation of changes in bioenergetic parameters of an eyestalk ablated intermoult prawn and their influences on tissue metabolism.

The size of a species is known to exert profound influence not only on the metabolic rate but also on the efficacy of metabolic adaptation to stress leading to size dependent differences in the adaptive ability of a species. Further various physiological and biochemical processes including metabolism and activity are found to be influenced by body size (Reddy and Davies, 1993). Hence body size is given due importance in this investigation.

The present investigation aims at investigating the influence of eyestalk ablation on moulting, bioenergetics and energy metabolism of *Macrobrachium malcolmsonii*.

In general allometric studies provide useful information regarding the size variations in a population and the relationship between different life history traits which could be used as indices to identify different life history stages of *M. Malcolmsonii* both for culture activities and experimental studies. Hence attempts were made to study allometry of *M. malcolmsonii* and results are presented in Chapter - I.

Moulting is responsible for growth in crustaceans. As such characterization of moult cycle stages in *M. malcolmsonii* could provide useful information on growth. Hence results on the effects of eyestalk ablation on moulting are presented in Chapter - II.

Bioenergetic studies are useful in understanding the energy budgets and physiological trade-offs of available energy between processes associated with growth. Thus the results pertaining to the effects of size and eyestalk ablation on bioenergetic variables are presented in Chapter - III.

Growth in crustaceans (or in any species) is associated with corresponding changes in the metabolism of total carbohydrate, lipid and protein. Determination of concentration of enzyme activities and metabolites relating to those major nutrient sources help understand the dynamics of growth in *M. malcolmsonii* in relation to size and eyestalk ablation. As such results concerning these aspects are presented in Chapter- IV.

The results obtained were summarized and appropriate conclusions drawn and presented under a Summary and Conclusions section.

It is humbly submitted that it is beyond the scope of the present work to identify the nature of influencing factor(s) present in the eyestalks and its mode of regulation. Nevertheless these investigations *per se* on *M. malcolmsonii* may be considered as a humble contribution to crustacean bioenergetics and physiology.

Introduction

Over the years crustaceans have received the bulk of experimental attention by reason of their easy availability. The most delicious invertebrates, at least judged by most humans, belong to the class Crustacea, considered as one of the most successful groups of aquatic animals since it includes a number of species with a wide range of geographical distribution (Vernberg and Vernberg, 1972). The order decapoda of the class Crustacea occupies an important position and includes commercially important prawns, shrimps, crabs, lobsters and crayfish. Among the commercially important crustaceans, prawns and shrimps undoubtedly occupy a pivotal position by virtue of their magnitude, fishery and food value.

Prawns, grouped into penaeid and non-penaeid species, have achieved considerable importance due to rapid growth and great potential in commercial market. Larger prawns mostly belonging to Penaeidae, Pandalidae, Hippolytidae, Sergestidae and Palaemonidae with considerably rich protein content are exploited for commercial purposes.

The fresh water prawn *Macrobrachium* is distributed throughout the tropical and subtropical zones of the world. There are about 150 species of *Macrobrachium* in the world (Brown, 1991) of which more than 50 species have been reported from India (Anonymous, 1993).

Holthuis (1980) has provided useful information on the distribution, habitat, local names and size ranges of commercial species of *Macrobrachium*. There is a wide interspecific variation in the size and growth rate of *Macrobrachium rosenbergii* (De Man), *Macrobrachium malcolmsonii (H. Milne Edwards)*, *Macrobrachium americanus* (Linnaeus) and *Macrobrachium carcinus* (Linnaeus) which probably are the largest species known amongst the genus *Macrobrachium*. Among the freshwater prawns of India, *M. rosenbergii, Macrobrachium idella* (Hilgendorf) and *Macrobrachium nobilli (*Henderson and Mathai*)* have been reported to be commercially important (Prakash and Agarwal, 1989; Jhingran, 1991) based on their global market value (Sampath Kumar et al. 2000). While disease resultant losses are a potential problem vis-a-vis the culture of brackish water species such as *Penaeus monodon* (Fabricius) and *Penaeus indicus (H. Milne Edwards)*, disease outbreaks have not been of significant concern as far as freshwater species such as *M. rosenbergii* and *M. malcolmsonii* are concerned.

M. rosenbergii is widely distributed in Indo-Pacific region while *M. malcolmsonii* is mainly confined to India, Bangladesh, Myanmar, Sri Lanka and Pakistan (New, 1988). *M. malcolmsonii* migrates upstream covering a distance of 1400Km where as *M. rosenbergii* prefers to stay in freshwater regions of about 200Km from an estuary (Kewalraman et al. 1971). Thus, *M. malcolmsonii* has great potential for recruitment as a competent species for culture in the upstream zones of rivers and streams. These are the two most suitable species for commercial farming in India and consequently the culture has extended to other countries also. Although a great deal of information is available on the culture aspects, physiology, biochemistry and endocrinology of *M. rosenbergii* (MacLean et al.

1994) relatively very little is known about the allometry, physiology, biochemistry and bioenergetics of *M. malcolmsonii*.

Macrobrachium malcolmsonii is prevalent in South-East Asia and is the most favoured species for commercial culture. *M. malcolmsonii* inhabits rivers, lakes, brackish waters and estuaries and is known as *"Tella royya"* especially in Andhra Pradesh. The distribution of palaemonid prawns in fresh and brackish waters from physiological point of view has been reviewed by Panikkar (1941). Even though they grow to a considerably large size, freshwater prawns of the genus *Macrobrachium* do not constitute a major fishery in the rivers (Bhimachar, 1962). But the riverine prawn production mainly contributed by *M. malcolmsonii* forms a sizable fishery only in river Godavari with an annual average of 85 tonnes representing 32.3% of the total fish yield (Jhingran, 1991). Langer and Somalingam (1993) obtained 128Kg/ha in 7 months, at Powarkheda (Madhya Pradesh) fish farm from monoculture system. Chandrasekharan and Sharma (1997) achieved a production of 157.5 Kg/ha along with 320Kg of fish in 6 months in polyculture. Knowledge on the prawn resources of other river systems of India is not very well known.

Among the freshwater prawns *M. malcolmsonii* and *M. rosenbergii* formed the cynosure of aquaculturists. Ever since indiscriminate commercial exploitation started, the numbers of the genus *Macrobrachium* started depleting to a very great extent. Thus there is an urgent need to overcome this problem and the best way to do this is to understand the physiology and biochemistry of this species vis-a-vis in relation to different environmental conditions.

An investigation into the biology of a commercially important species highlights its position and suitability for culture practices. Recent investigations on decapod crustaceans have attempted to predict growth including the onset of anatomical or functional sexual maturity (Hartnoll, 1973), sex ratio (Wenner, 1972), sex prediction (Allen, 1962), rate of growth (Tagatz, 1968), relative age of individuals in a given population (Hartnoll, 1974; Mauchline, 1976) and taxonomy (Williams et al. 1980). However no comprehensive investigation has been carried out on commercially important prawns of India with regard to these parameters except for studies on the biology of *Metapenaeus monoceros* (Fabricius) which include spawning and fecundity (Nalini, 1976) and some aspects of prawn fishery and age (Subramanyam, 1973; George, 1976). John (1957) studied the bionomics and life history of the prawn *M. rosenbergii* including fecundity, breeding potential and migration. Ling (1962) studied the life habits and culture of adult *M. rosenbergii*. Studies have also been conducted on the biology, development and larval stages of *M. rosenbergii* (Ling, 1969). Rajyalakshmi (1980) made some observations on the maturation and breeding of some estuarine palaemonid prawns and also conducted a comparative study of the biology of *M. malcolmsonii* and *M. rosenbergii* from the Hooghly and Godavari estuaries. Raman (1972) has extensively studied breeding, migration and feeding and fishery potential and biology of the giant fresh water prawns *M. rosenbergii* and *M. malcolmsonii* from southwest coast of India. Lio et al. (1969) studied the fecundity of this species in the laboratory while Rao (1958) has recorded the fecundity of allied species, *Metapenaeus affinis* (H.M.Ewards) and *Metapenaeus dobsonii* (Miers). Although polyculture of freshwater prawn *M.*

malcolmsonii with Indian and Chinese carps was reported by Reddy et al. (1985) literature on the biology of *M.malcolmsonii* is scanty.

The order Decapoda attracted the attention of very few investigators as far as the moult stages and setogenesis are concerned. Most investigators were primarily concerned with moulting and its staging. There is paucity of information on the chemical aspects of growth and development in crustaceans. In their review on "growth and development in crustaceans" Yamaoka and Scheer (1970) have given a detailed description of various stages of moult cycle. Ironically most of the work done so far in this field confined only to a few species of decapod crustaceans.

The life cycle of a typical decapod crustacean alternates between a relatively long intermoult period during which it feeds actively and a relatively short moult period during which it sheds the old exoskeleton and increase in size. The cycle is closely linked to the process of growth, as ecdysis is the only means by which a species with a rigid exoskeleton can grow. Since moult cycle is of prime consideration in most studies on crustacean physiology, it is necessary to establish a coherent and constant scheme describing various stages of moult sequence. Initially, Drach (1939) formulated certain characteristics for each and every moult stage of crustaceans as it differs from group to group and species to species. He was also the first investigator to describe stages of moult cycle by microscopic examination of the integument. He later modified the moulting cycle for natantians (Drach, 1944). Changes have also been suggested by Haiatt (1948) for grapsoid brachyurans and by Stevenson (1972) for macrura. Drach and Tchernigovtzeff (1967) and Mykles and Skinner

(1982) added some more information to the field of moult staging in different crustacean species.

Historically, studies on crustaceans have contributed considerably to the development of the concept of neurosecretion and neuroendocrine control. From the first discovery of a hormone in a crustacean in the 1920s to date, the field of crustacean endocrinology has undergone, as do many crustaceans during their development, a marked metamorphosis. The field has moved from the classical era of endocrinological tehcniques such as extirpation and additive methods, to the modern era of sophisticated biochemistry and molecular biology.

A neuroendocrine control mechanism for reproduction and development in crustaceans was demonstrated about 10 years after the first description of neurosecretion in crustacea. In 1943 Panouse found an ovarian inhibiting factor in the eyestalk as he observed that eyestalk removal from the females of *Palaemon serratus* (Pennant) during the period of genital rest leads to rapid increase in ovarian growth. Since that period the presence of so called gonad inhibiting hormone (GIH) in the X-organ-Sinus gland complex of the eyestalk has been confirmed in a number of crustaceans (Keller, 1992; Van Herp, 1998). Panouse's approaches are still applicable in aquaculture as unilateral eyestalk ablation is usually practiced to trigger moulting and vitellogenesis in shrimps.

Moulting in crustaceans refers to the periodic shedding of the old confining exoskeleton and the subsequent enlargement of the newly developed integument. It was regarded as a brief interruption in the normal life of the animal (Drach, 1939). Crustaceans pass through premoult (proecdysis), moult (ecdysis), postmoult (metecdysis) and intermoult (diecdysis) or "soft" and "hard" stages (Passano, 1960).

The modern view initiated by the studies of Drach (1944) and his coinvestigators (Lafon, 1948; Renaud, 1949) is that moulting in a decapod is not a physiologically restricted event interrupting the normal life but a process, which has great impact on the whole physiology. Metabolism, behaviour, reproduction and even the sensory acuity are all affected both directly and/or indirectly by periodic moultings.

Studies on moulting are facilitated by the ability to determine accurately each stage in the ecdysial cycle. Utilizing integumental changes, Drach (1939) first developed a criterion to divide the moult cycle of brachyurans like *Cancer pagurus* (Edwards) and *Maja squinado* (Herbst) into five major stages and 12 sub stages. Modification of Drach's original concept has allowed subsequent workers to delimit the moult stages of other decapods. The brachyuran (Drach-modified) and the corresponding natantian stages of intermoult cycle are described in Table 1 and 2 respectively. Histological changes (Travis, 1957), external features of the exoskeleton (Nagabhushanam and Vasantha, 1971) and morphological changes of the developing setae (i.e., setogenesis) (Freeman and Bartell, 1975) have been used as criteria for staging the ecdysial cycle. Setogenesis has been proved to be a very precise, quick and useful technique for moult staging, particularly in the diecdysic crustaceans which moult frequently (Knowels and Carlisle, 1956). In such individuals, the duration of post moult and intermoult stages is comparatively shorter than the premoult stage. Freeman and Bartell (1975) characterized the moult cycle of *Palaemonetes pugio* (Holthuis) (Table 3).

Table 1: Duration and characteristics of brachyuran intermoult stages (Passano, 1960)

Stage	Common name	Characteristics	Duration (% in total moult period)
A1	Freshly moulted	Water absorption and initial mineralization	0.5
A2	Soft	Exocuticle mineralization	1.5
B1	Paper shell	Endocuticle secretion	3.0
B2	----	Active endocuticle formation; hardening of chelae and tissue growth	5.0
C1	Hard	Main tissue growth	8.0
C2	----	Continued tissue growth	13.0
C3	----	Exoskeleton formation; occurrence of membranous layer	15.0
C4 (or)	----	"Intermoult"; organic reserves accumulation	30.1
C4 T	Permanent Anecdysis	Terminal Stage in certain species; No further growth	Permanent
D0	----	Epidermal and hepatopancreas activation	10.2
D1	----	Epicuticle and spine formation	5.0
D2	Peeler	Exocuticle secretion	5.0
D3	-----	Skeletal resorption	3.0
D4	About to moult	Ecdysial sutures open	1.0
E	Moult	Rapid water uptake and exuviations	0.5

Table 2 Duration and characteristics of natantian intermoult stages (Passano, 1960).

Stage	Characteristics	Duration (% in total moult period)
A1	Soft exoskeleton; matrix fills spines	2.5 (A1+A2)
A2	Branchiostegites flexible; retraction of spine matrix towards middle	
B1	Branchiostegites semi rigid; exoskeleton of a parchment consistency; spines with no conical base	16.5 (B1+B2)
B2	Partial formation of conical base	
C	Exoskeleton fully formed; conical base of spines complete; intermoult period	21.0
D0	Hormonal activation; epithelium separates from the old cuticle	Unknown
D1.1	New spine formation; nerve fibre retracts from the cavity of old spine	21.0
D1.2	New spine secretion	14.0
D1.3	Morphological details of new spines become visible	6.5
D2	Formation of pre-exuvial layers of new exoskeleton; resorption of old exoskeleton	17.0
D3	Old exoskeleton splits	1.0
E	Ecdysis	0.5

Eyes in decapod crustaceans are generally stalked and movable, and the eyestalk is known to contain a number of hormones or factors which apparently govern such diverse functions as moulting, growth, metabolism water balance, pigment dispersion and sexual maturity (Chang, 1985; Keller, 1992; Fingerman, 1997). Hence in order to understand the diverse roles of the eyestalk hormones, attempts have been made to deprive the crustaceans of their eyestalks (Unilateral/ bilateral eyestalk ablation technique; Bray and Lawrence, 1992; Treece and Fox, 1993; Fox and Treece, 2000).

Table 3: Characterization of the moult cycle of *Palaemonetes pugio* (Freeman and Bartell, 1975).

Stage	Characteristics	Duration of each stage(days)
Postmoult (metecdysis)	Presence of tissue filled setae; no internal cones; rostrum easily bent and the body soft	1.0
Intermoult (diecdysis)	Appearance of internal cones; base of setae alone are tissue filled; rostrum firm and exoskeleton rigid	1-2
Premoult (proecdysis)	Retraction of epidermis parallel to the base of each setae	1-2
D0	Epidermis retracts further; primordia of setal developing areas visible feebly	1-2
D1-1	Widening of setal forming areas	1-2
D1-2	Setal forming areas widen to their greatest width near the epidermis; new setae become visible.	2-3
D2	New setae clearly seen from epidermis to the old setae; epidermal surface regresses further from exoskeleton	3-7

Several methods have been adopted by different researchers to accomplish destalking. Caillouet (1973) performed ablation by cutting the eyestalk near its base with a pair of sharp scissors, the wound was cauterised immediately with a pencil type soldering iron to avoid the loss of hemolymph. Lumare (1979) did not cauterise the wound inflicted due to ablation. Primavera (1985) incised the eyeball with a sharp blade, allowed the fluid to ooze out and squeezed the contents of the eyeball outwards between the thumb and the forefinger and crushed the eyestalk two to three times to destroy the tissue. Rodriquez (1979) simply squeezed the contents of the eyeball out and crushed the eyestalk by pressing between the fingers. Muthu and Laxminarayana (1980) destalked the penaeid prawns with medical electrocautery apparatus used in surgery. Thus cutting the eyestalk and sealing the wound simultaneously ensured 100% survival of the destalked animals. Subjecting *P. monodon* to three different ablation methods, Makinouchi and Primavera (1987) found that electrocautery and ligating were more effective than pinching.

The efficacy of eyestalk ablation in triggering precocious moulting has been well documented (Skinner, 1984; Rotllant et al. 2000). The chronological development of this subject is quite interesting because the effects of eyestalk removal on moulting were known long before they were fully appreciated. Perhaps the first few observations in this connection were those of Zeleny (1905) and Megusar (1912). Subsequently several authors have studied eyestalk ablation vis-a-vis moulting in *Homarus americanus* (H. M. Edwards) (Aiken & Waddy, 1976; Mauviot and Castell, 1976), *Panulirus argus* (Latreille) (Quackenbush and Herrnkind, 1981); *Panulirus homarus* (Linnaeus) (Radhakrishnan and Vijayakumaran, 1984a), *P. pugio* (Freeman and Bartell, 1975), *Palaemon elegans* (Rathke) (Webster, 1985), *M. nobilli* (Pandian and Sindhu Kumari, 1985), *P. monodon* (Poernomo & Hamami, 1983), *Procambarus clarkii* (Girard) (Nakatani & Otsu, 1979), *Uca pugilator* (Rathbun) (Abramowitz and Abramowitz, 1940; Guy Selman, 1953) *Macrobrachium lamarrei* (H.M.Edwards) (Marian et al. 1986), *Homarus americanus*

(H.M.Edwards) (Flint, 1972), *P.argus* (Travis, 1954), *Panulirus cygnus* (George) (Dall, 1977) and *Macrobrachium lanchesteri* (De Man) (Ponnuchamy et al. 1981). These observations show that reports on eyestalk ablation experiments are inconsistent.

In several species of lobsters (*H.americanus*), prawns (*M. nobilli*; *M. lamarrei*), cray fishes (*P.clarkii*) and crabs (*U. pugilator*) eyestalk ablation induced precocious moulting while the same technique failed to evoke response when performed in other species or repeated even in the same species (Quackenbush and Herrnkind, 1981). Sochasky et al. (1973) attributed Travis's (1954) failure in inducing precocious moulting in *P. argus* to gonadotropic interference, as the lobsters used by her were prepubertal or mature. Quackenbush and Herrnkind (1981) contradicted the results of Travis (1954) and concluded that eyestalk ablation does accelerate precocious moulting as well as gonadal activity but not simultaneously. This could be due to the fact that moulting and reproduction are antagonistic events in these lobsters. Thus long term experiments are necessary to arrive at a definite conclusion. Flint (1972) showed an extension of the intermoult period in *H.americanus* which could be due to the fact that Flint did not fix the stage of the moult cycle and eyestalk ablation was performed ignoring the fact that the technique is effective only when performed at the appropriate stage of the moult cycle (Freeman and Bartell, 1975). Further, Ponnuchamy et al. (1981) while reporting no increment in moulting frequency in eyestalk ablated *M. lanchesteri* opined that a longer experimental duration may yield significant results.

Barring comparatively fewer number of contrasting results obtained in *H. americanus* (Aiken & Waddy, 1976; Mauviot and

Castell, 1976), *P, argues* (Quackenbush and Herrnkind, 1981), *P. homarus* (Radhakrishnan and Vijayakumaran, 1984a), *P. pugio* (Freeman and Bartell, 1975), *P. elegans* (Webster, 1985), *M. nobilli* (Pandian and Sindhu Kumari, 1985), *P. monodon* (Poernomo & Hamami, 1983), *P. clarkii* (Nakatani & Otsu, 1979), *U. pugilator* (Abramowitz and Abramowitz, 1940; Guy Selman, 1953) and *M. lamarrei* (Marian et al. 1986) it can be said with a fair amount of certainty that the factor removed along with the eyestalk is moult inhibitory in nature. Hanstorm (1931) rediscovered the presence of Bellonci's (1882) x-organ in the eyestalks. Moult inhibiting hormone (MIH) is secreted by the X-organ present in the eyestalk and stored in the sinus gland. A moult-inhibiting factor, indole alkylamine-like substance was isolated from the eyestalk of the prawn *Pandalus jordani* (Rathbun) by Soyez and Kleinholz (1977). Presence of MIH in the eyestalk has been confirmed further by the injection of eyestalk extract into eyestalk ablated individuals which ultimately postponed the ensuing moult. Freeman and Bartell (1975) and Bruce and Chang (1984) thus demonstrated the presence of MIH in the eyestalks of *P. pugio* and *H.americanus,* respectively. Eyestalk removal results in decreased titre of MIH in the hemolymph (Mattson, 1986). On the otherhand, ecdysteroid (the moulting hormone) titres synthesized in and secreted by the Y-organ (Gabe, 1953) which is situated in the head region of the cephalothorax were found to increase (Chang, 1985; Mattson and Spaziani, 1986; Karplus and Hulata, 1995). Thus moulting is facilitated. The route taken by 20-OH-ecdysone as it activates moulting in insects has been mapped (Bonner, 1982) and it is assumed that a similar route is being followed in crustaceans also. It has recently been shown that an exuviation factor (Charmantier - Daures and

Vernet, 1974; Charmantier - Daures and De Reggi, 1980) from the Y organ also plays a significant role in moulting process.

High mortality among eyestalk ablated crustaceans is not uncommon. Brown and Cunningham (1939) reported 100% mortality before, during or shortly after a single moult in eyestalk ablated crayfish *Cambarus clarkii* (Girard). Thus the gain made through faster growth was offset by poor survival. Smith (1940) and Mauviot and Castell (1976) emphasised the role of diet on the survival of ablated decapods. Further eyestalk ablated *Cambarus clarkii* were fed "liberally" to improve survival and to moult several times. Impact of dietary constituents on growth and survival of eyestalk ablated crustaceans has been reviewed by New (1976). Basic nutritional studies which measure metabolic functions after eyestalk ablation are thus warranted.

Eyestalk ablation is capable of not only inducing precocious moulting but also enhancing growth (Silas, 1982; Pandian and Sindhu Kumari, 1985; Radhakrishnan and Vijayakumaran, 1984a). However some authors (Nakatani and Otsu, 1981; Karplus and Hulata, 1995) reported that there was no actual increase in growth due to eyestalk ablation. They argued that the increase in weight observed in freshly moulted eyestalk ablated animals was due to the abnormal intake of water. It can be pointed out that though there is water uptake during late premoult or postmoult, this water is replaced by tissue growth at a later stage. Nakatani and Otsu (1981) tested the eyestalk ablated crayfish, *P.clarkii* with an intermittent feeding regimen. As such studies considering dry weight of animals at an appropriate stage of the moult cycle with appropriate feeding regimen are necessary to authentically state whether eyestalk ablation enhances growth or not. Bilateral extirpation of stalked eyes not only resulted in accelerated moult but also has been shown to delay metamorphosis in larvae by the production of extra larval stages in the shrimp, *Palaemon macrodactylus* (Rathbun) (Little, 1969) and *Palaemonetes varians* (Leach) (Le Roux, 1984); the lobster *H.americanus* (Charmainter et al. 1988); and intermediate stages in crabs, *Rhithropanopeus harrissi* (Gould) (Costlow, 1968) and *Sesarma reticulatum* (Say) (Costlow, 1966; Freeman and Costlow, 1980). The endocrine control of morphogenesis in crustaceans is not well understood but evidence suggests that inhibitory factors from the sinus gland in the eyestalks are involved.

Eyestalk ablation in adult crustaceans results in premature gonadal development irrespective of the breeding season. An understanding of the factors which regulate the reproductive cycle of a species of commercial interest is an important prerequisite for programming its culture (Lumare, 1979).

While investigating the gonad stimulating factor in the central nervous system of crustaceans, Otsu (1963) found that implantation of pieces of thoracic ganglion into immature and sexually inactive females of the freshwater crab *Potamon dehaani* (Rathbun) resulted in considerable ovarian growth. Gomez (1965) reported that the brain, besides thoracic ganglion, contains ovary stimulating hormone in the crab *Paratelphusa hydrodromous* (Herbst). Gomez (1965), Adiyodi and Adiyodi (1970), Hinsh and Bennet (1979), Sarojini et al. (1982) and Adiyodi (1985) attributed maturation of ovary to the gonad stimulating hormone (GSH). Eastman - Reks and Fingerman (1984) pointed out that thoracic ganglion extracts from female *U. pugilator* induced precocious ovarian maturation in intact and eyestalkless crabs. Inhibitory action of eyestalks on ovarian development has been reported for various crustacean species by several authors (Sarojini et al. 1996, 1997; Vaca and Alfaro, 2000).

Application of this method (destalking) in laboratory culture of prawns was first initiated by Idyll (1971) and followed by Arnstein and Beard (1975), Emmerson (1980) and others. Initial mortality was high after eyestalk ablation in almost all experiments and those which survived initially had well developed gonads but due to regression ovaries died later. Duronslet et al. (1975) observed that oocyctes from ablated animals demonstrated normal growth but did not under go meiosis. Alikunhi et al. (1975) reported successful spawning of bilaterally eyestalk ablated *Penaeus merguiensis* (de Man) and *P. monodon*.

The high mortality and inability of females to spawn after bilateral eyestalk ablation prompted scientists to abandon this method. Arnstein and Beard (1975) found that maximum ovarian development and survival could be realised only through unilateral eyestalk ablation in *Penaeus orientalis* (Kishinouye), *Penaeus occidentalis* (Street) and *P. monodon*. Since then unilateral eyestalk ablation has been employed in different penaeid prawns by several authors (Browdy and Samocha, 1985; Makinouchi and Primavera, 1987).

In many of decapods, removal of eyestalk accelerated moulting but failed to induce ovarian maturation (Passano, 1953). Sagi et al. (1997) and Rotllant et al. (2000) found that moulting and reproduction are antagonistic in adult brachyurans. During eyestalk removal whatever be the set of process (moulting or reproduction) in progress, the dominant becomes apparent.

In several species, eyestalk ablation results not in moulting but in premature yolk deposition in the ovary, both during the non breeding and breeding seasons. In certain species like *P. hydrodromous* (Gomez, 1965) and *Scylla serrata* (Forskal) (Rangneker and Deshmukh, 1968), destalking induces yolk deposition even in pre-pubertal stages. Eyestalk removal in young or non-breeding adult males has been claimed to induce precocious spermatogenesis, enlargement of vasadeferentia (Gomez, 1965) and hypertrophy and hypersecretion in the androgenic glands (Khalaila et al. 1999).

The effects of eyestalk removal are not uniform in different species of decapoda. Various factors such as eyestalk principle, age, food, stress (due to over crowding, frequent holding, poor water quality etc.,), salinity, pH, temperature and light seem to influence the course of events that follow eyestalk ablation (Muthu and Laxminarayana, 1982). Experiments involving eyestalk ablation have revealed that in some phases of the ovarian cycle, the operation fails to accelerate vitellogenesis (Adiyodi, 1968; Anilkumar and Adiyodi, 1985) and that the production and release of ovary inhibiting hormone by eyestalk and susceptibility of ovaries to this factor appears to be cyclic in crustaceans.

In reptantians like crabs various aspects of reproduction such as egg maturation, ovulation (release of egg from the follicles), spawning (oviposition) and incubation of eggs are completed during the intermoult stage (Adiyodi and Adiyodi, 1970; Chang, 1995). Spawning follows ecdysis in natantians (Charniaux - Cotton, 1985). Somatic and reproductive growth occurs simultaneously in natantian decapods (Sarojini et al. 1983). Pillay and Nair (1970) and Thomas (1974) described the annual reproductive and nutritional cycles of certain penaeid prawns.

In caridean prawns such as *Palaemon serratus* (Panouse, 1947), *Palaemon paucidens* (DE Hann) (Kamiguchi, 1971) and *M. nobilli* (Pandian and Balasundaram, 1982) spawning is obligatorily preceded by a moult. Synchronous occurrence of moulting and reproduction demands apportioning of available energy for both the processes. Rotllant et al. (2000) observed stimulation of gonads in adults by increasing the concentration of methylfarnesoate and ecdysteriods in the ablated spider crab, *Libinia emarginata* (Leach). Millamena and Quinito (2000) reported that the ablated female mud crab, *S. serrata* fed with formulated diet in the presence of essential nutrients and vitamins showed enhanced production of good quality eggs. Sagi et al. (1997) reported that the unilateral eyestalk ablation significantly increased spawning activity in young females over already spawned females. Junio- Menez and Ruinata (1996) considered unilateral eyestalk ablation as a viable option in accelerating growth and food conversion efficiency and enhance the period of survival in spiny lobsters, *Panulirus ornatus* (Fabricius) *to* marketable size. Chen et al. (1995) reported that the eyestalk ablation in crayfish, *P. clarkii,* shortened moult interval and induced mortality suggesting that temperature and other factors also might play an important role in mortality.

Chaves (2000) observed that removal of eyestalk resulted in acceleration of ovarian development. Daily injection of four sinus gland equivalents reduced ovarian growth of ablated females by about 50%. Koshio et al. (1992a) attributed the growth resulting from eyestalk ablation in juvenile H. *americanus* to more efficient energy utilization. Mohamed and Diwan (1991) observed the precocious maturation of ovary in uni- and bilaterally eyestalk ablated Indian white prawn, *P. indicus*. Besides, bilaterally ablated prawns were shown to exhibit abnormal behaviour along with moulting and reproduction simultaneously.

Bioenergetics:

The study of energy changes accompanying biochemical reactions in living cells is termed "Bioenergetics" or "Biochemical thermodynamics" (Sibly and Calow, 1986). Bioenergetics became an important branch of Life Sciences to study the energetic values in living animals (Collins and Anderson, 1999). The study of bioenergetics explains the production and loss of energy by the living organisms to different life activities, plays a significant role in understanding the mechanism of nutrition, digestion, locomotion, respiration, excretion, reproduction, growth and development of living organisms and throws light on the energy budget of an individual as well as a population (Whiteledge et al. 1998).

Energy is often defined as the capacity to do work. Food provides an animal with two vital commodities – chemicals and energy. If the animal is growing, some energy will be retained in the chemical bonds of growth materials. Some energy will presumably be passed onto the environment as heat through work done by the animal. In addition energy will be lost in the chemical bond of nitrogenous wastes (U) and in faeces (F). The following equation represents the energy budget of an animal by Klekowsky and Duncan (1975).

$$C = (F + U + R_{rout} + ASDA + SFG)$$

C: Food consumption (Energy consumption); F: Energy lost through faeces, U: Energy lost through excretion, R_{rout}: Energy allocated to metabolism, ASDA: Energy allocated to apparent specific dynamic action, SFG: Scope for growth.

The measurement and analysis of energy used by organisms for their existence and growth is of great interest in ecology. Energy cannot be measured directly. One can only measure the transformation of energy from one form to another. The officially recommended unit for energy in all its manifestations is now *'Joule'*, but *'calorie' is* likely to persist, as its use has been so wide - spread. [(1 calorie = 4.2 Joules); Elliott (1976)].

Partitioning of energy between metabolic demands likely has profound influence on the ecological success (fitness) of animals. Changes in energy allocation (amount/pattern) can be used as indicators of environmental stress - external constraint limiting the rates of resource acquisition, growth or reproduction (Grime, 1989).

Energy budgets serve as a powerful framework for identifying the most important aspects in the life of an individual. Bioenergetic models related to energy intake, expenditure and productivity have been established for several fish species. Brett (1971) established bioenergetic model for *Oncorhynchus nerka* (Walbaum), Adams et al. (1982) and Rice and Cochran (1984) for *Micropterus salmoides*

(Lacepede) and Diana (1983) and Bartell et al. (1986) for *Esox lucius* (Linnaeus). The use of bioenergetics models is an economical, reasonably accurate and feasible method for stimulating growth or estimating ration for a variety of fish (Kitchell et al. 1977; Stewart et al. 1983; Rice and Cochran, 1984; Stewart and Binkowski, 1986).

The process of feeding relies upon morphological, physiological and behavioural mechanisms that allow animals to acquire and ingest food items, digest complex food stuffs to their simpler component molecules and absorb nutrients contained in food (Sibly, 1981).

Ting (1970), Liao and Huang (1975), Liao and Murai (1986) and Lei et al. (1989) in *Penaeus monodon* (Fabricius), Liu (1983) and Chen et al. (1991) in *Penaeus chinensis* (Osbeck) and Dalavia (1986) in *Penaeus japonicus* (Bate) have demonstrated the effects on oxygen consumption of both extrinsic factors such as water temperature, salinity, dissolved oxygen and photoperiod and intrinsic factors such as body weight. Stephenson and Knight (1980) showed a decline in oxygen consumption of *M. rosenbergii* with increasing salinity from freshwater to 28%. In *Macrobrachium acanthurus* (Weigman) oxygen consumption rates increased proportionally with temperature and were inversely proportional to biomass whereas respiratory rates decreased with increase in salinity indicating altered metabolic costs (Gasca-Leyva et al. 1991). Nelson et al. (1977) and Clifford and Brick (1978) found no significant relationship between the amount of food consumed and specific dynamic action in a fresh water prawn, *M. rosenbergii*. Feeding was found to double respiratory rate in *P. japonicus* (Sacyanan and Hirata, 1986).

Marangos et al. (1990) reported that ammonia excretion rates were five times higher in post-larvae than in adults of *P. japonicus*. Chen and Kou (1991) reported that *P. japonicus* subadults exposed to increased levels of ambient ammonia had significantly higher hemolymph ammonia levels after 2 hr. It was also found that oxygen consumption and ammonia excretion in adolescent *P. japonicus* increased with increase in ambient ammonia indicating the potential effects of external factors on energy loss in respiration and / or excretion (Chen and Lai, 1992). In the face of changing environmental conditions the ability to regulate oxygen consumption has been found to be developed to varying extent in decapod crustaceans (Mangum et al. 1973; Herreid, 1980).

Availability of food is a critical environmental variable. A change in food supply would be reflected in a number of dependent variables like growth, exuviation, egg production (size and number of eggs) etc. As moulting is an integral phenomenon in the life of a decapod a portion of the assimilated energy has to be allocated for structural (as exuvium) as well as functional (as metabolic) costs of moulting. While the former amounts to 1.1 KJ, the latter amounts to 4.7 KJ in Juvenile *M. rosenbergii*. The interrelationships between food intake, assimilation efficiency and reproduction have not been examined systematically thus far in *M.malcolmsonii*. Feeding trials with different restricted rations will help understand the strategy; a species adopts to adjust itself to the oscillating food availability in the habitat.

Carbohydrate metabolism:

Existence of glycolytic and pentose phosphate pathways in various crustacean species has been confirmed by many workers (Huggins and Munday, 1968; Hohnke and Scheer, 1970; Sadok et al. 1997).

The TCA cycle coupled with electron transport system and oxidative phosphorylation and adenosine diphosphate (ADP) has also been demonstrated in many crustacean species (Huggins and Munday, 1968; Chen and Lehninger, 1973). TCA cycle is the main pathway through which the various food stuffs namely carbohydrates, proteins and lipids get oxidised completely into CO_2 and water by a series of biochemical reactions. Presence of TCA cycle intermediates and related enzymes has been demonstrated in crustaceans (Bellany, 1962; Huggins, 1966; Vosloo et al. 1996). Besides glycolysis and TCA cycle, several alternate pathways for the carbohydrate metabolism have also been reported in crustacean species (Hohnke and Scheer, 1970; Chang and O' Connor, 1983; Castro et al. 1998).

That the carbohydrate metabolism is under the control of eyestalk hormones was first demonstrated by Abramowitz et al. (1944). Since then, changes in carbohydrate metabolism after eyestalk ablation or after administration of crude eyestalk extract to the eyestalk ablated crustaceans have been reported by many workers (Keller, 1966; Bauchau et al. 1968; Dircksen, 1992). It is well documented that eyestalk ablation causes variations in hemolymph sugars in crustaceans (Keller and Sedlmeier, 1988). However these variations in the hemolymph sugars are not uniform in different species. Eyestalk ablation has been shown to induce hyperglycemia in some species such as crayfish *Orconectes virilis* (Hagen) (McWhinne and Saller, 1960), *Ocypode plantitaris* (Fabricius) (Parvathy, 1972) and *Paratelphusa jacquemontil* (Rathbun) (Rangneker et al. 1971) and hypoglycemia in some other species such as *Metapenaeus monoceros* (Rangneker & Madhyastha, 1971), *P. monodon* (Kuo et al. 1995), *M. rosenbergii* (Lin et al. 1998), *P. indicus, M. monoceros* (Kishori et al. 2001) and *Scylla serrata* (Reddy and Kishori, 2001). Thus although much evidence favours hyperglycemic nature of functioning, there are reports equally favouring hypoglycemic regulation. This confirms that the effects of eyestalk ablation are not uniform, but differs in crustaceans, indicating species specific responses.

Protein metabolism:

Proteins are the ubiquitous macromolecules in a biological system and constitute about one fifth of an animal's body on wet weight basis (Swaminathan, 1983) and 68 - 85% on dry weight basis (Jauncey, 1982). They not only serve as building blocks but also as fuel to yield energy (Pandian, 1989). Functionally, proteins exhibit a great diversity and constitute a heterogenous group having diverse physiological functions such as the formation of structural and contractile elements, hormones, catalysts, toxins, protective agents etc. (Lehninger, 1984). The concen-tration of proteins in a tissue is the balance between the rate of synthesis and degradation or catabolism (Grained and Seglen, 1981; Tavil and Cooksley, 1983).

Nitrogen metabolism is both a pre-condition to and a consequence of the crustaceans' nutrition, growth, development, energetics, and physiological adjustment to various endogenous and exogenous variables (Claybrook, 1983).

Amino acids formed on protein degradation serve as the precursors of protein synthesis and gluconeogenesis and are the currency through which protein metabolism operates (Nelson and Cox, 2000). Generally invertebrates have a higher pool of intracellular free amino acids than the vertebrates. Arthropods have the highest amino acid concentration (Florkin, 1966). Available data largely deals with amino acid composition of the crustacean muscle (Schoffeniels and Gilles, 1970). Considerable variations do exist in the distribution of amino acids among different species. For example glutamic acid concentration is very high in *Carcinus meanas, Eriocheir sinensis, Maja squinado* and *Homarus vulgaris* and much lower in *Calanus finmarchinucs, Cancer pagurus* and *Nepunus pelagicus* (Schoffeniels and Gilles, 1970). In addition, a considerable variation exists in the same tissue of different species. There is also differential distribution of free amino acids in different tissues of a single species (Schoffeniels and Gilles, 1970).

The existence of most enzymes concerned with amino acid metabolism has been identified in most crustaceans (Huggins and Munday, 1968; Florkin and Schoffeniels, 1969). It is demonstrated that the crustacean tissues have both anabolic and catabolic enzymes concerned with the amino acid metabolism (Schoffeniels and Gilles, 1970). The deamination of amino acids is reported to be the prime source in crustacean tissues for the ammonotelic pattern (Vernberg, 1983).

Neuroendocrine regulation of nitrogen metabolism in crustaceans is also well studied (Neiland and Scheer, 1953; Thornborough, 1968; McWhinnie and Mohrherr, 1970). Neiland and

Scheer (1953) reported that eyestalk principle controls RNA synthesis and RNAase activity in crustaceans. Destalking in crab, *O. senex senex* leads to an increase in proteins and this was evidenced through increased incorporation of C^{14} amino acids into proteins (Ramamurthi et al. 1981). A similar trend was reported in other crustaceans (Skinner 1965; McWhinne and Mohrherr, 1970). Total nitrogen was found to decrease in crayfish (Fingerman et al. 1967) following eyestalk ablation. Decrease in proteins following eyestalk ablation in immature *S. serrata* and restoration upon eyestalk extract injection was also observed (Radhika et al. 1988).

Lipid Metabolism:

Lipids along with carbohydrates and proteins are important as food for many animals. Metabolism of lipids embraces free fatty acids, sterols and other complex lipids (Hoar, 1984). Total lipids are of particular importance in the metabolism, because they present the bulk of stored energy. Fats occur subcutaneously and are stored in the adipose tissues (Lehninger, 1984).

The principal storage site for lipids in crustaceans is the hepatopancreas (Midgut gland / digestive gland). It is more analogous to the liver of vertebrate in function (Gilbert and O'Connor, 1970; Chang and O' Connor, 1983). More than 90% of neutral lipid represents triglycerides (O' Connor and Gilbert, 1968).

During the moult cycle hepatopancreatic lipids along with body lipids are differentially utilized as energy source and for the synthesis of certain hormonal as well as structural components of the cells (Gilbert and O'Connor, 1970). In some cases lipids are sequestered at one stage, while at another stage they are utilized as energy sources when the animal almost solely dependent on lipids for its energy (Gilbert and O'Connor, 1970). Eyestalk ablation led to a decrease of total hemolymph lipids in *Chasmagnathus granulata* and of free fatty acids in *Carcinus meanas* (Santos et al. 1997).

Neurosecretory hormones released by crustacean eyestalk play a prominent role in the metabolic regulation also and this is revealed through a number of distinctive biochemical characteristics. These concern changes in the body protein, lipid and carbohydrate composition and control of water, calcium and phosphate alterations (Travis, 1957).

The incorporation of fatty acids as fat depots in hepatopancreas was found to increase markedly during premoult stage. Thus during premoult, the decapod crustaceans might have enhanced the capacity to store ingested lipids and to synthesize lipids from the catabolic products of ingested carbohydrates (Chang and O'Connor, 1983). Once the fatty acids are synthesized, they are rapidly incorporated through synthetic routes leading to glycerides, phospholipids, and sterol esters etc. (Gilbert and O' Connor, 1970). The role of sterols in arthropods is multifold. Sterols are found to act as precursors for synthesis of the moulting hormone (Gilbert, 1969) and also as components for sub cellular membrane synthesis (Gilbert and O' Connor, 1970). In arthropods neuroendocrine hormones / principles were found to regulate metabolism and metabolic pathways by exerting influence on the concerned enzyme systems (Raghavaiah et al. 1980). Bollenbacher et al. (1972) and Chang and O' Connor (1983) reported lipid biosynthesis in most crustaceans following destalking. On the contrary inhibition of lipid biosynthesis was also reported in the land crab *Gecarcinus lateralis* and crayfish (O'Connor and Gilbert, 1969). Increase in lipid biosynthesis following eyestalk ablation appears to be independent of the changes in the level of moulting hormone (Bollenbacher et al. 1972).

Statement of the Problem:

It has been demonstrated that the eyestalks of prawns and other crustaceans possess hormones or factors that are capable of regulating different aspects of metabolism (Sehnal, 1971). The role of neuroendocrine system in moulting, growth and reproduction in crustaceans received good coverage (Sagi et al. 1997; Rotllant et al. 2000). The effects of neuro-secretary hormones on carbohydrate (Lin et al. 1998; kishori et al. 2001) protein (Raghavaiah et al. 1980; Ramamurthi et al. 1981) and nitrogen metabolism (Radhika et al. 1988) were found to be different from one species to another.

Investigations Into the allometry and biology of commercially important prawns highlight their suitability for aquaculture. Hence attempts were made to study the biology of *M. malcolmsonii* in I chapter. Moult cycle characterization and effect of eyestalk ablation on moult cycle duration have been dealt with in the II chapter. Results on bioenergetics variables in relation to eyestalk ablation have been included in the III chapter. Changes in carbohydrate, protein and lipid metabolism as a result of eyestalk ablation were covered in the IV chapter.

Main objective of the present work:

Though a great deal of information is available on the physiological and metabolic effects of eyestalk ablation in various crustaceans, attempts have not been made to investigate the effects of eyestalk ablation on moulting, bioenergetics and metabolism in *M. malcolmsonii* a species with great aquaculture potential. The present work is a humble effort in the direction of understanding the effects of eyestalk ablation on diverse physiological functions of *M. malcolmsonii*. This by no means, is a comprehensive piece of work vis-a-vis the problem chosen but every possible effort was made to collect relevant data; to present meaningful interpretations of the results obtained and to arrive at definitive conclusions.

Materials and Methods

Experimental animal

The palaemonid prawn, *Macrobrachium malcolmsonii* (H. M. EDWARDS, 1844) occurs abundantly in freshwater lakes, reservoirs and rivers in the plains throughout India along with *Macrobrachium rosenbergii* (DE MAN, 1879). The juveniles of these prawns enter from the near by estuarine waters into freshwater during monsoon rains, grow and attain maturity.

Macrobrachium malcolmsonii (Figure1) which has high commercial value and economic importance is called as *"Tella royya"* in Andhra Pradesh. Because of its distribution both in freshwater and brackish water bodies M. *malcolmsonii* constitute an important link in the food web of estuaries. While the distribution of *M. rosenbergii* is restricted mostly to freshwater regions of upto 200 kms from the estuary (Rajyalakshmi, 1968; Kewalraman et al. 1971), *M. malcolmsonii* migrate upstreams covering a distance of 1400 kms (Rajyalakshmi, 1968; Ahmed, 1984).

Recent statistics show that *M. malcolmsonii* constitutes a major fishery in river Godavari with an annual average landing of 85 tonnes representing 32% of the total fish yield (Jhingran, 1991). Because of its great commercial value, abundant availability and easy adaptability to the laboratory conditions, *M. malcolmsonii* is selected as the experimental animal in the present investigation.

Biology of the test species:

Macrobrachium malcolmsonii which is comparatively smaller in size, better taste and more meat than *M. rosenbergii* is considered to be an ideal candidate for culture. *M. malcolmsonii* can attain a length of 230 mm and compete with *M. rosenbergii* in several respects. The farming potential of any species is assessed by taking into account various factors viz., growth, distribution, feeding habit, prolonged breeding, fairly good fecundity, availability of adequate feed, good market etc. *M. malcolmsonii* possess all the above qualities. It is found to be most suitable for aquaculture on account of fast growth rate in freshwater as well as in low saline waters. It has omnivorous feeding habit, hardy in nature to tolerate low temperature, has dissolved oxygen compatibility for polyculture, and has resistance against several diseases, lower cannibalistic tendencies and good market value.

Classification:

Phylum -	Arthropoda
Class -	Crustacea
Sub class -	Malacostraca
Super order -	Eucarida
Order -	Decapoda
Suborder -	Natantia (Boas, 1880)
Tribe -	Caridea (Dana, 1852)
Family -	Palaemonidae (Samovelli, 1819)
Sub family -	Palaemoninae (Dana, 1852)
Genus -	*Macrobrachium* (Bate, 1868)
Species -	*malcolmsonii* (H. Milne Edwards, 1844)

Figure 1: *Macrobrachium malcolmsonii (H.Milne Edw. 1844)*

Salient features of *Macrobrachium malcolmsonii (H.Milne Edw.)*

1. Carop-propodal articulation of pereopod occur only at one fixed point, pleopods are well developed and modified for swimming.

2. Pereopod III never chelate, pleura of second abdominal segment noticeably overlaps those of the first segment. Abdomen is usually short and curved.

3. Pleurobranchiae are present at the base of the maxilliped III. Posterior margin of telson has two pairs of spines and two or more setae.

4. Mandibles are usually with incisorial process or maxilliped III without flate and foliate.

5. Dactyl of last three pereopods simple.

6. Basal crest distinctly elevated provided with 5-9 teeth. In younger specimens, the II leg has the palm swollen and the fingers larger than the palms: Carpus of II leg in adult male shorter than chela.

Procurement and maintenance:

Macrobrachium malcolmsonii were collected from *Penna River* near Nellore, located 134 km north-east of Tirupati, Andhra Pradesh, India (Figure 2), using a dragnet, locally called as *"Kontivala"* (a velon screen net of 3x2 mts, length having 2.5/1 cm mesh size) and transported late in the evening (in a sheltered truck) to the laboratory at Tirupati in plastic containers containing filtered river water. During transportation water temperature was maintained at 26 - 28°C with the addition of cool filtered river water from a storage tank (200 L earthen vertical tank wrapped in wet gunny bag). Water in the plastic containers was continuously aerated using battery-operated aerators with biomass water volume maintained at 1 g/L. As a result mortality due to handling, shock and hypoxia was successfully brought down to 5 - 10%.

Figure 2: Map showing the Penna River Drainage

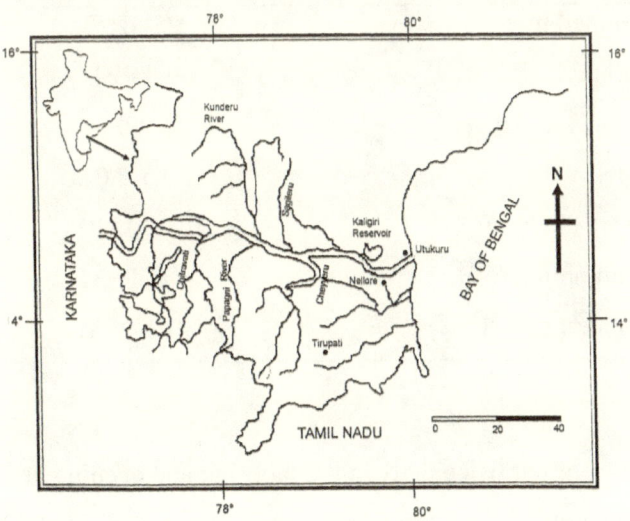

In the laboratory, prawns exhibiting symptoms of disease or retarded activity or moulting are discarded. Healthy and active prawns were transferred to clean plastic holding tanks filled with uncontaminated aerated tap water at 26-28^{0}C. As the laboratory conditions are much different from those of their natural habitat, all the prawns were allowed to acclimate to the new environmental conditions for about seven days. Since dense aggregations cause mass mortality (Muthu, 1981) prawns were divided taking size into consideration into batches of 15 each and kept in 15 L holding tanks / troughs (Figure 3). Water in the tanks was aerated to have 75-80% aeration. Water quality parameters (Table 4) were analysed following the standard methods (AOAC, 1981) before stocking and also during the experimental period.

Table 4: Water quality parameters and their optimal levels used in the experiments

S.No.	Parameter	Level
1	Temperature	25-32^{0}C
2	Salinity	0.5 PPT
3	Dissolved oxygen	5-7 PPM
4	pH	7.5-8.0
5	Total alkalinity	50-100 PPM
6	Ammonium (NH_4^+)	<15 PPM
7	Ammonia (NH_3)	<1.0 PPM

Prawns were fed twice daily in the morning and evening at 5% body weight *(ad libitum)*, on commercial feed (pellet code no.9004) obtained from CP Aquaculture (India) pvt. Ltd., Chennai.

The unconsumed food, faecal matter, exuvia and dead prawns, if any, were removed every time before feeding and water in each trough was replaced daily with fresh, filtered tap water for better survival of the prawns (Muthu, 1981).

Figure 3: Indoor rearing facility for *Macrobrachium malcolmsonii*

Experimental design:

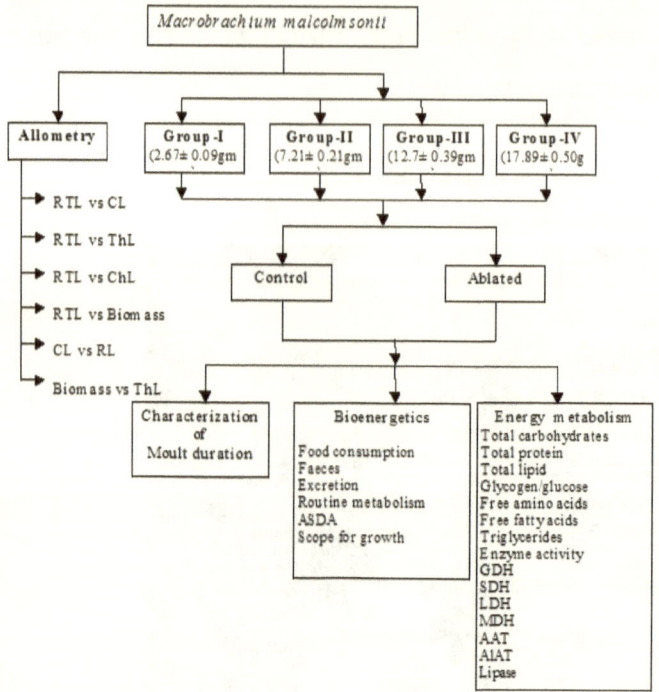

Allometry:

60 individuals of *M. malcolmsonii* were randomly selected from the main stock and the wet biomass (gm) of each individual was determined using a sartorius balance to nearest gm (0.1 gm). Length measurements of the following body parts were taken to the nearest mm (0.1 mm).

RTL - Rostral telson length: From the tip of the rostrum to the tip of the telson.

RL - Rostral length: Dorsally from the tip of the rostrum to the point from where the eyestalks arise.

CL - Carapace length: Dorsally from the point where the eyestalks emerge, to the posterior end of carapace.

ThL - Thoracic region length: Dorsolaterally from the tip of the rostrum to the posterior end of carapace.

ChL - Chelate leg length: From the tip of dactylus to the coxal end.

W - Biomass: Weight of the individual.

Relationships between RTL Vs CL/ThL/ChL/biomass, CL Vs RL and W Vs ThL have been examined through a regression model. The explanatory variable (X) and its influence on different dependent variables (Y) have been studied. This simple regression of Y on X has been studied through curvefitting with the help of EXCEL software (Microsoft Office XP).

Characterization of Moult Stages:

Four size groups (2.67±0.09gm; 7.21±0.21gm; 12.7±0.39gm; 17.89±0.50 gm) (Figure 4) of *M.malcolmsonii* were separated from the main stock. Six individuals were selected from each size group and maintained in separate plastic containers keeping all the water quality variables and feeding schedules at an optimal level as explained earlier for observations of setal development. Observation of setal development in crustaceans such as prawns which possess thin and transparent cuticle is relatively easy compared to those with a thick and opaque integument (Drach, 1944; Kurup, 1964; Stevenson, 1972; Rao et al. 1973).

In order to delineate the moulting stages in *M.malcolmsonii*, microscopic examination of dorsal surface of endopodite / exopodite of uropod was carried out by the method of Drach and Tchernigovtzeff

(1967). From the days of first moult, prawns were kept under observation till they complete atleast two moult cycles. The uropods of *M. malcolmsonii* were removed with the use of a fine dissecting scissors. The tip of the pleopods were immediately wet mounted on a microscope slide in physiological saline, covered with a coverslip and examined with light microscope at 10X and photographed.

Figure 4: *Macrobrachium malcolmsonii* of four different size groups selected for the study.

Moult stage duration:

More than 20 prawns of early premoult stage of (D_0) were isolated from the total animal collection and kept separately in the laboratory for observing the time duration from D_0 stage to succeeding stages. Every 24hr, the uropods of these animals were observed individually. The time duration of each substage of premoult stage (D_0, D_1 etc.,) was determined by the extent of epidermal retraction and setal development until the ecdysis stage (E). Similarly, the time duration of postmoult and intermoult stages were determined by hardness of the exoskeleton and the uropodal changes. Unilateral eyestalk ablation was performed on intermoult prawns of each size group by cutting the eyestalk at the base with fine scissors without prior ligation, and cauterizing the wound after ablation (Chakravarty, 1992).

Bioenergetic studies:

The conceptual framework (Figure 5) of energy allocation in the control and experimental prawns was described through an energetic equation proposed by Klekowsky and Duncan (1975).

Figure 5: Conceptual framework of the energy allocation model

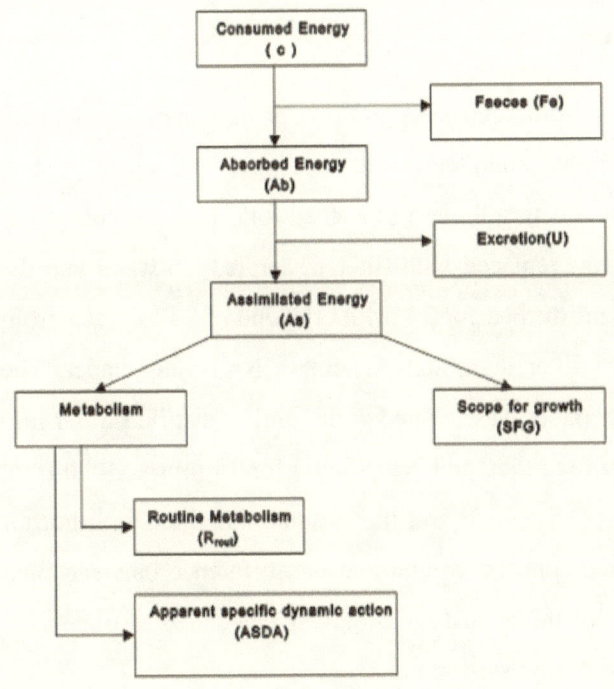

$C = (F + U + R_{rout} + ASDA + SFG)$

C: Food consumption (Energy consumption); F: Energy lost through faeces, U: Energy lost through excretion, R_{rout}: Energy allocated to metabolism, ASDA: Energy allocated to apparent specific dynamic action, SFG: Scope for growth.

Measurement of food consumption (C):

Pelleted feed equivalent to 5% of prawn biomass (Ci) was weighed to the nearest milligram in an electric balance and provided to the individual prawn. Individual prawns (n=6 each for control and eyestalk ablated groups) were fed separately in plastic containers. After 2-3 hr of feeding, left over feed was collected, blotted dry and dried in a hot air oven at 80° C and weighed (Cf). The difference between Cf and Ci was taken as the amount of food consumed (C; gm feed) by each prawn.

Estimation of faeces (Fe):

Six prawns from each size group kept in separate containers immediately after the completion of feeding were used in this experiment. After collecting the left over feed following feeding, water in each container was replaced with filtered, aerated tap water and the prawns were left undisturbed for 24 hr. At the end of 24 hr water from each container was filtered through Whatman No.1 filter paper. The residue on the filter paper was washed with 10ml of distilled water into a preweighed centrifuge tube and centrifuged for 15 min at 4000 rpm. The supernatant was discarded and the residue was dried in a hot air oven at 80±10C to a constant weight and the difference between final and initial weights of the centrifuge tube was calculated to obtain the amount of faeces and expressed in mg.

Estimation of ammonia (U):

Ammonia excretion (mg NH_3/24hr) was measured individually following Solorzano (1969). To 50ml of water sample, 2 ml phenol solution, 2 ml sodium nitroprusside solution, and 5 ml of oxidizing reagent were added and mixed thoroughly after each addition. The colour is allowed to develop at room temperature for 1 hr and the absorbance is recorded at 640 nm against reagent blank. Six prawns from each group were starved (unfed) for one day before commencing the experiment and maintained individually in one litre of filtered, aerated tap water. After 24 hr, the amount of ammonia excreted was measured taking 50 ml water sample and expressed as endogenous ammonia excretion (mg NH_3/L/24 hr). Later the prawns were fed as explained earlier. After collecting the left over feed following feeding, water was replaced with filtered, aerated tap water and maintained for a period of further 24hrs. Later the water samples were collected for the estimation of ammonia as explained above. The difference between the amount of ammonia excreted for 24 hr by the fed and unfed prawns was calculated and represented as the amount of exogenous ammonia excreted. Since no detectable quantity of urea excretion was found in prawns as noted in some other few organisms (Regnault, 1987) the amount of ammonia excreted alone has been considered as nitrogenous excretion.

Routine respiratory metabolism (R_{rout}):

Oxygen consumption of *M. malcolmsonii* with random movement was measured using a flow-through respirometer coupled with dissolved oxygen electrode (ELICO) (Davies et al. 1992) and expressed as ml O_2 /24hr. *Macrobrachium malcolmsonii* were found not to exhibit complete resting for a considerable period of time during the experimentation. Hence oxygen consumption of randomly moving unfed prawns was considered as routine respiratory metabolism (R_{rout}).

Description of the apparatus:

A wider and larger glass tube (length: 20 Cm ; diameter: 7 cm) closed at one end and fitted with a rubber cork at the other end was used as the respiration chamber in place of the glass syringe used by Davies et al. (1992) (Figure 6). This chamber contains two small glass tubes at two subterminal ends for attachment of rubber tubes as shown in the diagram. A 15 L glass aquarium filled with filtered tap water was used as the reservoir. Dissolved oxygen concentrations were maintained in the reservoir by bubbling compressed air through air stones.

Figure 6: Schematic diagram of the flow-through respirometer

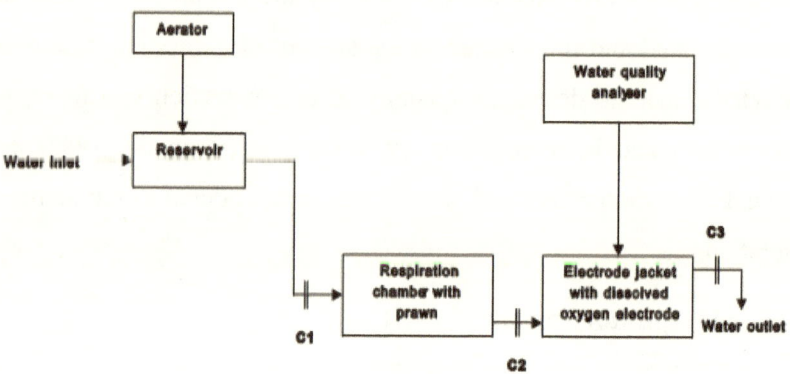

The proximal end of the respiration chamber is connected to the reservoir through a rubber tube (inlet) and provided with a clamp C1while the distal end of the respiration chamber is connected to the lower portion of an electrode jacket through another rubber tube (outlet) and provided with clamp C2. An electrode jacket is a conical flask provided with one holed rubber cark and two side tubes-one at the neck region and the other at the bottom region. The electrode jacket accommodates the dissolved oxygen electrode, (Model PE-130; ELICO). Dissolved oxygen electrode connected to a Water Quality Analyser (Model PE-136; ELICO) measure the amounts of dissolved oxygen (DO) present in the water. Water from the electrode jacket is flushed out through the tube located at the neck of the electrode jacket and fitted with clamp C3. All the connections were made airtight by applying Vaseline, and the water flow is maintained by operating clamps C1, C2 and C3.

Measurement of dissolved oxygen:

Initially C1 and C2 were opened to draw filtered water into respiration chamber and electrode jacket as well. Clamp C3 was opened after the respiration chamber and electrode jackets were completely filled with water. All the clamps were then adjusted in such a way that the flow rate of water is maintained at 1 ml/min without air bubbles. The amount of oxygen present in the water of electrode jacket was measured once every five minutes for one hr (initial) with the help of the DO electrode and water quality analyser. Then the cork of the respiration chamber was removed and unfed (starved for 1 day) prawn was introduced into the respiration chamber and the cork was closed tightly. Now for every five minutes the amount of dissolved oxygen in the water of electrode jacket was measured for one hr (final). Similar measurements were taken for six replicates of both control and ablated prawns of each size. The average amount of dissolved oxygen in initial and final samples were calculated separately and the difference between the mean of initial and final values was represented as the amount of oxygen consumed by each prawn for 1 hr. Similar calculation has been carried out for each size group separately.

Amount of oxygen consumed (ml O_2) = IO_2 - FO_2

$IO2$ = Amount of dissolved oxygen present in the initial sample.
$FO2$ = Amount of dissolved oxygen present in the final sample.

Apparent specific dynamic action (ASDA):

Apparent specific dynamic action [(ASDA) (mlO_2/day) which includes energy costs of excitement, activity, digestion, absorption and biochemical transformation of absorbed food (Beamish, 1974; Tandler and Beamish, 1981) was measured using a flow through respirometer coupled with dissolved oxygen electrode (ELICO) (modified version of Davies et al. 1992). Before taking measurements each unfed prawn was acclimated separately in the respiration chamber for 1 hr to reduce the excitement, if any, due to the stress of handling. After the prawn settled down to the normal condition the amount of oxygen consumed was measured once every 2 hr for 24 hr continuously. The total amount of oxygen consumed for 24 hr (ml O_2) was calculated and considered as control value (R_{rout} unfed). Subsequently feed pellets weighing approximately 5% of the prawn body weight were introduced into the respiration chamber. Measurements of oxygen consumption were taken continuously when the prawn exhibited exploratory behaviour and started consuming the feed. Measurements of oxygen consumption were taken till the level decreased to that of the control prawn. Total amount of oxygen (mlO_2) consumed by fed prawn was calculated and considered as experimental value (R_{rout} fed). ASDA was calculated as the difference between the amount of oxygen consumed by fed and unfed *M.malcolmsonii* and the average of six individuals was expressed as mlO_2.

$$ASDA\ (mlO_2) = R_{rout}\ fed - R_{rout}\ unfed$$

Rrout fed = Routine respiration of fed prawns
Rrout unfed = Routine respiration of unfed prawns

Determination of Calorific values:

Food and faeces:

Samples of food and faeces were dried to a constant weight in a hot air oven at $80\pm1^{0}C$ for 24 hr. The dried samples were used for energy estimations. If not used immediately the same were stored in a desiccator for further analysis. Calorific content of dried samples was determined in a semi-microbomb calorimeter (Toshniwal DT 100) using the standard procedure described in the instruction manual. For every measurement food consumption and egestion were estimated in terms of dry weight and then converted into energy units and represented in J/24hr.

Excretion:

Values of ammonia excretion (mg NH_3) were converted to energy equivalents using a co-efficient of 20.4 J/mg NH_3 (Elliott and Davison, 1975).

Metabolism:

Energy costs of routine metabolism (R_{rout}) and ASDA were calculated using the oxycalorific value of 20.2J / mlO_2 (Elliott and Davison, 1975).

Parameters calculated:

Rate of food consumption (Rc):

Rate of consumption was calculated as follows

$$\text{Rate of consumption} = \frac{\text{Amount of energy consumed (J)}}{\text{Body biomass (gm)}}$$

Absorbed energy (Ab):

Absorbed energy was measured using the measurements of consumed energy (C) and energy lost in faeces (Fe).

Hence,
$$Ab = C - Fe$$

Absorption efficiency (AbE):

Based on the values of absorbed energy (Ab) and consumed energy (C), the absorption efficiency (AbE) was calculated using the following formula.

$$AbE\ (\%) = \frac{C - Fe}{C} \times 100$$

Assimilated energy (As):
$$As = C - (Fe + U)$$

Assimilated energy was calculated based on the measurements of consumed energy (C), energy lost in faeces (Fe) and the energy lost in excretion (U). The following equation was used to calculate assimilated energy.

Assimilation efficiency (AsE):

$$AsE = \frac{C - (Fe + U)}{(C - Fe)} \times 100$$

Based on the measurements consumed energy (C), energy loss in egestion (Fe), and energy lost in excretion (U) assimilation efficiency was calculated as shown below.

Scope for growth (SFG):

Scope for growth (SFG) is the potential energy available for an organism that could be channelled into growth once the metabolic requirement has been met. SFG is calculated as the difference in energy content of ingested food (C) and the total energy loss in faeces (Fe) and excretion (U) and energy used for routine respiratory metabolism (R_{rout}), apparent specific dynamic action (ASDA).

$$SFG = C - (F + U + R\ rout + ASDA)$$

Energy metabolism:

All biochemical estimations and enzyme activity levels were measured in muscle and hepatopancreas of 6 individuals of each size group of both control and eyestalk ablated prawns. Concentrations of glucose, total protein and total lipid were estimated in freshly collected samples of hemolymph.

Hemolymph collection:

Hemolymph was collected from the joint between II and III pleopods using 1ml hypodermic syringe with a 25 gauge needle and rinsed with EDTA buffer (anti coagulant) (Lee et al.1999).

Biochemical estimations:

Total Carbohydrate (TCHO):

Total carbohydrate content was estimated by the method of Carrol et al. (1956). 5% homogenate of muscle and 2% homogenate of hepatopancreas were prepared in 10% TCA. The homogenates were centrifuged at 1000g for 15 min at 4^0C. To 0.2 ml of TCA supernatant, 4 ml anthrone reagent was added and boiled for 15 min. The colour was read at 620 nm in a spectrophotometer (Systronics UV-VIS Spectrophotometer 108) using blank consisting of 0.2 ml TCA and 4 ml anthrone reagent. Total carbohydrate content was expressed as mg/gm wet weight.

Glycogen:

Glycogen content was estimated by the method of Carrol et al. (1956). 10% homogenates of hepatopancreas and muscle were prepared separately in 5% TCA and centrifuged at 1000g for 15 min. To one volume of TCA supernatant, 5 volumes of 95% ethanol was added and allowed to stand overnight in cold (4^0C). After complete precipitation, the contents were centrifuged again for 15 min at 1000g. The supernatants were discarded and the residue was dried by placing the tubes in an inverted position for 10min at room temperature. The residue was then dissolved in 1 ml of distilled water. A reagent blank (with 1 ml of distilled water) was also prepared and 5 ml of anthrone reagent was added to each test tube by constant stirring. The tubes were kept in a boiling water bath for 15 min and allowed to cool to room temperature. The colour was read at 620nm in a spectrophotometer (Systronics UV-VIS Spectrophotometer 108) against blank. The values were expressed as mg glycogen / gm wet weight.

Hemolymph glucose:

Hemolymph glucose content was determined by the method of Kemp and Mayers (1954). To 1 ml of hemolymph, 1 ml 5% trichloroacetic acid containing 0.1% silver sulphate was added to deproteinise the sample. The contents were centrifuged for 10 min at 1000g. To 0.5 ml of the supernatant 4.5ml analar sulphuric acid was added. The mixture was heated for 6 minutes in a boiling water bath and the contents were cooled to room temperature. The colour was read at 520nm against a reagent blank consisting of 1.5ml of 5% TCA and 4.5ml of analar sulphuric acid using spectrophotometer (Systronics UV-VIS Spectrophotometer 108). The values were expressed in mg glucose / 100ml hemolymph.

Total protein:

Total protein was estimated by the method of Lowry et al. (1951). Hepatopancreas and muscle tissues were quickly isolated and 2% homogenates were prepared in 10% TCA and centrifuged at 600g for 15min. The residue was dissolved in appropriate quantities of 1N sodium hydroxide. To 0.5 ml of the above solution 4ml of copper reagent was added. The samples were allowed to stand for 10 min, at the end of which 0.4ml of Folin phenol reagent was added. Finally the optical density of the colour developed was measured using spectrophotometer (Systronics UV-VIS Spectro-photometer 108) at 600nm against a reagent blank. Values are expressed in mg protein / g wet weight.

Total protein estimation was similarly carriedout in hemolymph. To 0.5ml of hemolymph 0.5 ml of 10% TCA was added in order to precipitate the proteins and other steps followed were similar as described above. Values are expressed as mg protein/ml hemolymph.

Free amino acids:

Total free amino acids in the tissues and hemolymph of prawn were estimated by the method of Moore and Stein (1954). 2%

homogenates of hepatopancreas and muscle were prepared separately in 10% TCA and centrifuged at 600g for 15min. To 0.5ml of the supernatant, 2ml Ninhydrin reagent was added, boiled in a water bath for 5min and cooled immediately to room temperature. The contents were made upto 10ml with distilled water and the colour was read at 570nm in a spectrophotometer (Systronics UV-VIS Spectrophotometer 108) against a blank. The blank consists of 0.5ml 10% TCA and 2ml ninhydrin reagent.

To estimate the amount free amino acids present in the hemolymph, 0.5ml of hemolymph was taken 0.5ml of 10% TCA was added to it and centrifuged. Other steps followed were similar as described above. Tyrosine was used for standard graph preparation. The free amino acid content is expressed as μmoles of tyrosine equivalents / gm wet biomass or μmoles of tyrosine equivalents / 1 ml hemolymph.

Total lipid:

Total lipid was estimated by the method of Folch et al. (1957). Muscle and hepatopancreas were isolated, weighed to the nearest mg, kept at 60^0C in a hot air oven for one day and dry weights were taken. A known amount of dry, powdered tissue was homogenized in chloroform: methanol (2:1) mixture (16ml). The contents were placed in 25ml capacity test tubes and kept in a water bath, set at 61^0C. The contents were allowed to boil for 5 min and cooled to room temperature. The volume was made upto 25ml with chloroform, 10ml normal saline (0.9%) was added and shaken vigorously. The lower aliquot was collected into a preweighed beaker and kept in a hot air oven at 60°C for evaporation. The beaker was cooled and again weighed. The weight of the residue in the beaker is equal to the amount of lipid present in the sample. To estimate the total lipid present in the hemolymph, 0.5 ml of hemolymph was mixed with chloroform: methanol (2:1) and centrifuged at 1000 g for 5 min and the rest of the procedure is same as detailed above. Total lipid content is expressed in mg lipid / gm of wet biomass or mg lipid/ ml hemolymph.

Triglycerides:

Triglycerides were estimated by the method of Natelson (1971) with slight modifications as given below. Triglycerides were assayed by hydrolysing them to glycerol and by estimating the liberated glycerol. An aliquot of 4 ml from the total lipid extract was collected and 200 mg of dried salicylic acid was added. The contents were vigorously shaken and supernatant was evaporated to dryness. To each tube, while it was still hot, 0.4 ml absolute ethanol plus 0.1 ml alkaline barium solution were added and heated for 30 min at 80EC to ensure complete hydrolysis. The reagent blank was prepared by taking 0.4ml ethanol and 0.1 ml alkaline barium solution.

After heating the tubes were cooled, 1ml 2N sulphuric acid was added and shaken well. To this 0.1ml sodium periodate solution was added and mixed well. The contents were kept aside for 10 min and 0.1ml sodium arsenate solution followed by 5 ml chromotrophic acid reagent were added and the contents were heated. The colour developed was read at 575 nm in a spectrophotometer (Systronics UV-VIS Spectrophotometer 108) against reagent blank. Standard graph was prepared by taking various concentrations of glycerol. The triglyceride content was expressed as mg glycerol/gm wet biomass or mg glycerol/ml hemolymph.

Free fatty acids:

Free fatty acid content was estimated by the method of Natelson (1971). After determining the total lipid content, 2ml of 95% ethanol was added to the lipid residue followed by a drop of 0.1% phenolphthalein (prepared in pure ethanol). The contents were titrated against 0.02N potassium hydroxide. Development of pink colour is the end point.

Similarly, blank (2ml ethanol) was also titrated against 0.02N potassium hydroxide.

Net titre value X 0.02 = milli equivalent (m.eq.) free fatty acid.

Milli equivalent (m.eq.) fatty acid X 227 = Free fatty acid content in mg.

Since 1 ml of 0.02N KOH – 0.02 m.eq. of free fatty acid and 227 is an assumed average mol. weight for the fatty acid).

Free fatty acid content is expressed as mg/gm wet biomass or mg/ml hemolymph.

Enzyme activity:

Succinate dehydrogenase (SDH): (Succinate - oxidoreductase; E.C. 1.3.99.1)

Succinate dehydrogenase activity was estimated separately in the muscle and hepatopancreas by the method of Nachlas et al. (1960). 5% homogenate was prepared for each tissue in 0.25 M ice cold sucrose solution and centrifuged at 2000g for 15min. The supernatants were taken for enzyme assay. The reaction mixture in the final volume of 2.0ml contained: 40 μmoles of sodium succinate, 100 μmoles of phosphate buffer (pH 7.4), and 4 μmoles of INT (2-(4-Iodopheny1) 3-(4-Nitrophenyl) - (5- phenyl tetrazolium chloride). The reaction was initiated by the addition of 0.5 ml supernatant. The reaction mixture was incubated at 37°C for 30min in a thermostatic water bath. After incubation the reaction was arrested by adding 5ml glacial acetic acid and the formazan formed due to reduction of the dye was extracted overnight in 5ml toluene at 5^0C. The colour was read in a spectrophotometer (Systronics UV-VIS Spectrophotometer 108) at 495nm against a blank (5ml toluene). The enzyme activity was expressed in μmoles formazan formed/mg protein / hr.

Lactate dehydrogenase (LDH) :(L - Lactate - NAD oxidoreductase; E.C. 1.1.1.27)

Lactate dehydrogenase activity was estimated separately in the muscle and hepatopancreas by the method of Srikantan and Krishna Murthy (1955). 5% homogenate was prepared for each tissue in 0.25 M ice cold sucrose solution and centrifuged at 2000g for 15min. The supernatant was used for the enzyme assay. The reaction mixture in a final volume of 2.0ml contained: 100 μmoles of phosphate buffer (PH 7.2), 0.1 μmole of NAD, 40 μmoles of sodium lactate and 4 μmoles of INT. The reaction was initiated by the addition of 0.5 ml supernatant. The reaction mixture was incubated at 37^0C for 30min in a thermostatic water bath. After incubation the reaction was arrested by adding 5ml glacial acetic acid and the formazan formed due to reduction of the dye was extracted overnight in 5ml toluene at 5^0C. The colour was read in a spectrophotometer (Systronics UV-VIS Spectrophotometer 108) at 495nm against a blank (5ml toluene). The enzyme activity was expressed in μmoles formazan formed/mg protein / hr.

Malate dehydrogenase (MDH): (L- Malate - NAD oxidoreductase; E.C. 1.1.1.37)

Malate dehydrogenase activity was estimated separately in the muscle and hepatopancreas by the method of Nachlas et al (1960). 5% homogenate was prepared for each tissue in 0.25 M ice cold sucrose solution and centrifuged at 2000g for 15min and the supernantant was used for the enzyme assay. The reaction mixture in a final volume of 2.0ml contained: 100 μmoles of phosphate buffer (pH 7.2), 0.1 μmole of NAD, 40 μmoles of sodium malate and 4 μmoles of INT. The reaction was initiated by the addition of 0.5 ml supernatant. The reaction mixture was incubated at 37^0C for 30min in a thermostatic water bath. After incubation the reaction was arrested by adding 5ml glacial acetic acid and the formazan formed due to reduction of the dye was extracted overnight in 5ml toluene at 5^0C. The colour was read in a spectrophotometer (Systronics UV-VIS Spectrophotometer 108) at 495nm against a blank (5ml toluene). The enzyme activity was expressed in μmoles formazan formed/mg protein/hr.

Glutamate dehydrogenase (GDH): (L - Glutamate - NAD oxidoreductase; E.C. 1.4.1.3)

Glutamate dehydrogenase activity was estimated separately in the muscle and hepatopancreas by the method of Lee and Lardy (1965). 5% homogenate was prepared for each tissue in 0.25 M ice cold sucrose solution and centrifuged at 2000g for 15min and the supernatant was used for the enzyme assay. The reaction mixture in a final volume of 2.0ml contained: 100 µmoles of phosphate buffer (pH 7.2), 0.1 µmole of NAD, 40 µmoles of 0.1 M sodium glutamate (pH 7.4) 4 µmoles of INT. The reaction was initiated by the addition of 0.5 ml supernatant. The reaction mixture was incubated at 37^0C for 30min in a thermostatic water bath. After incubation, the reaction was arrested by adding 5ml glacial acetic acid and the formazan formed due to reduction of the dye was extracted overnight in 5ml toluene at 5^0C. The colour was read in a spectrophotometer (Systronics UV-VIS Spectrophotometer 108) at 495nm against a blank (5ml toluene). The enzyme activity was expressed in µmoles formazan formed/mg protein/hr.

Protease activity:

Protease activity was estimated separately in the muscle and hepatopancreas by the method of Moore and Stein (1954). 5% homogenate was prepared for each tissue in 0.25 M ice cold sucrose solution and centrifuged at 2000g for 15min and the supernatant was used for the enzyme assay. The reaction mixture in a final volume of 2.0ml contained: 100 µmoles of phosphate buffer (pH 7.0), 20 mg of denatured hemoglobin and 1.0 ml of the supernatant. The reaction mixture was incubated at 37EC for 30min in a thermostatic water bath. After incubation the reaction was arrested by adding 2 ml 10% TCA. Unincubated samples were treated with 2 ml of 10% TCA prior to the addition of the enzyme source. After some time the contents of both the incubated and unincubated samples were filtered. To 0.5 ml of the aliquot, 2 ml of Ninhyrin reagent is added and kept in boiling water bath for 5 min and then cooled. The volume is made upto 10 ml with distilled water and the colour was read in a spectrophotometer at 570 nm against a blank. All the samples were corrected for zero time controls. The enzyme activity was expressed in μmoles tyrosine equivalents / mg protein/ hr.

Alanine aminotransferase (AlAT): (DL - Alanine - 2 -oxoglutarate aminotransferase; E.C. 2.6.1.2)

AlAT activity was assayed separately in muscle and hepatopancreas by the method of Reitman and Frankel (1957). 2% homogenate was prepared for each tissue in ice cold 0.25M sucrose solution and centrifuged at 1500g for 15min. The supernatant was collected for the enzyme assay. The reaction mixture in a final volume of 1ml contained: 40 μmoles of DL - Alanine (pH 7.4), 2 μmoles of α-ketoglutarate, 100 μmoles of phosphate buffer (pH 7.4) The reaction was initiated by the addition of 0.4 ml of enzyme source. The reaction mixture was incubated at 37EC for 20min. The reaction was arrested by adding 1ml of 2,4 - dintrophenyl hydrazine solution and allowed to stand at room temperature for 20min. Zero time controls were maintained for all the samples by the addition of 1.0 ml of ketone reagent prior to the addition of the enzyme source. Colour developed, due to the addition of 10ml of 0.4N sodium hydroxide, was read at 545nm in a spectrophotometer against reagent blank. The Enzyme activity was expressed as μmoles of pyruvate formed / mg protein / hr.

Aspartate aminotransferase (AAT): (L- Aspartate -2- oxaloglutarate aminotransferase; E.C. 2.6.1.1):

AAT activity was assayed separately in muscle and hepatopancreas by the method of Reitman & Frankel (1957). 2% homogenate was prepared for each tissue in ice cold 0.25M sucrose solution and centrifuged at 1500g for 15min. The supernatant was collected for the enzyme assay. The reaction mixture in a final volume of 1ml contained: 40 µmoles of L-Aaspartate, (pH 7.4), 2 µmoles of α-ketoglutarate, 100 µmoles of phosphate buffer (pH 7.4). The reaction was initiated by the addition of 0.1 ml of enzyme source. The reaction mixture was incubated at 37^0C for 20min. The reaction was arrested by adding 1ml of 2,4 - dintrophenyl hydrazine solution and allowed to stand at room temperature for 20min. Zero time controls were maintained for all the samples by the addition of 1.0 ml of ketone reagent prior to the addition of the enzyme source. Colour developed; due to the addition of 10ml of 0.4N sodium hydroxide was read at 545nm in a spectrophotometer (Systronics UV-VIS Spectrophotometer 108) against reagent blank. The enzyme activity was expressed as µmoles of pyruvate formed/mg protein/hr.

Lipase activity:

Lipase activity in the selected tissues of control and experimental prawns was assayed by the method of Colowick and Kaplan (1955). 5% homogenate was prepared for each tissue in ice cold distilled water and centrifuged at 1000g for 5min. The supernatant was collected for the enzyme assay. The reaction mixture consists of 0.5ml of substrate (Tween - 20), 1ml of buffer, 0.1ml of phenol Red, 0.4ml of distilled water, and 1ml of the supernatant. The enzyme activity was initiated by adding 1ml of supernatant and incubated at 37^0C for 15min. The enzyme activity was arrested by heat denaturation. To this, 5ml of Folch mixture (3 parts chloroform; 1 part methanol) is added and vigorously shaken and kept for 30min and titrated against 0.02N NaOH till the appearance of pink colour, which denotes the end point. A blank prepared in the same manner using heat killed supernatant was also titrated against 0.02N sodium hydroxide. 1ml of 0.02N NaOH is equivalent to 100 lipase units. The enzyme activity is expressed in mg free fatty acid liberated/mg protein/ hr.

Sham Controls

Apart from the usual controls, a set of sham controls was also run along with the experimentals, particularly for the eyestalk ablated prawns. The sham for destalking consists of just injuring one of the eyestalks. The effect of such operations was found to be ineffective.

Statistical Analysis:

For each parameter, the mean of six individual observations (for both control and experimental groups) were taken into consideration. Statistical significance of the data was analysed through two way ANOVA (Analysis of variance); SNK (Student - Newman - Keuls) test and regression analysis (Zar, 1984).

Chapter 1: Allometry

In 1917 D'Arcy Wentworth Thompson published a treatise on *"Growth and Form"* in which he argued that mathematical and physical laws not only underlie biological form but they actually generate biological form (Gould, 1992). Although such a heterodox view has not overtaken biology, Thompson's visual approach proved aesthetically compelling. In particular, his use of Cartesian coordinates to analyze changes in animal form influenced generations of biologists. It is hard to imagine any biologist of the past eighty five years who has not seen a reproduction of one of Thompson's "transformed fish". It is worth reflecting on what Thompson and Huxley were trying to do. Both distilled the obvious complexity of variations in organismic form in the simple pattern, in an attempt to identify what they considered the essence of the problem. Huxley's approach of measuring the relative sizes of organs typically along a single dimension such as length was arguably more analytical than Thompson's approach. Huxeley found that simple models of growth provided good fits to bivariate plots of these linear data and thus believed that he had identified laws of growth. The field has, however, failed to progress further. It seems likely that we cannot infer the mechanisms that control relative growth from the patterns of differential growth, and that we require a mechanistic understanding of growth that is only now becoming accessible. The mechanistic study of relative growth is essentially a new field with little data. Static allometry is the scaling of relationship among individuals between one organ and total body size or between two organs, after growth has been ceased or at a single developmental state. The study of allometry originated as a problem in morphology.

Almost any feature of an organism (eg. metabolic rate) can be compared to size to reveal possible functional relationships (Schmidt - Nielsen, 1984; Reiss, 1989). Huxley (1932) switched easily between ontogenetic and static allometry, seeing them as two parts of single problem. But here attention was focused on the problem of static allometry, or how different sized individuals produce adult organs that scale approximately to body size. Scatter diagrams related to the above relations between the various parameters together with observed values were presented in figures 1.1, 1.2, 1.3, 1.4, 1.5 and 1.6.

The size range of 60 specimens of *M. malcolmsonii* examined was widely spread. The smallest *M. malcolmsonii* measured 48mm and the largest measured 130mm rostral telson length (RTL). Similarly whole biomass ranged from 0.87 to 26.78g.

A large number of individuals encompassing a wide size range occur in *M. malcolmsonii*. However, the measurements obtained appeared generally similar to those which might have attained from a consecutive moulting progression. *M. malcolmsonii* consisted of a number of individuals which can be ranked more or less regularly with increase in size. Several interesting allometric relations have been obtained for *M.malcolmsonii*.

RTL vs Biomass:

The regression is a good fit and on an average the biomass increased by 0.274 g whenever RTL increases by one cm. In this model the intercept is negative (-15.512) which means that there is a minimum value of RTL at which the biomass will be positive (Figure 1.1). The correlation between RTL and biomass is 0.930. The biomass increment is linear thus demonstrating positive allometry. This clearly denotes that growth in *M. malcolmsonii* is regular and the biomass increment is a function of length may be due to the extensive change of both morphological features.

Figure 1.1: Allometric relation between RTL and Biomass in *M.malcolmsonii*

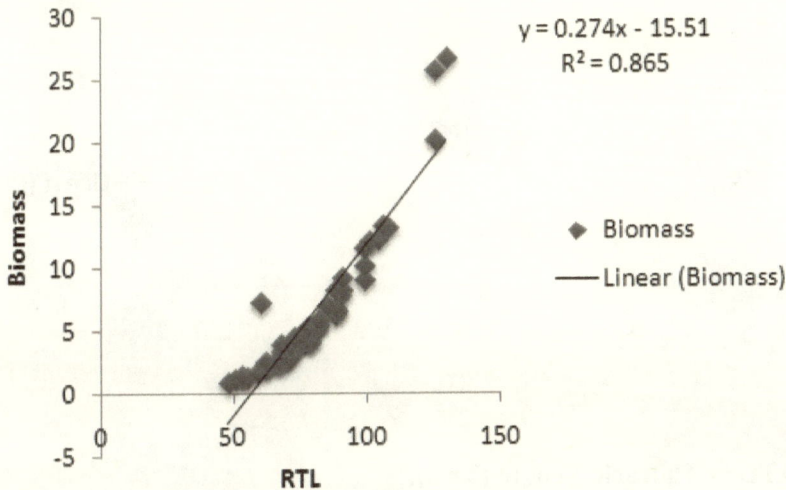

RTL vs Carapace Length (CL):

The regression is a good fit and on an average the CL increased by 0.285 cm whenever RTL increases by one cm. In this model the intercept is negative (-2.796) which means that there is a minimum value of RTL at which the CL will be positive (Figure1.2). The correlation between RTL and CL is 0.867. The RTL is positively allometric to CL. Higher slope value (0.284) clearly demonstrate the contribution of carapace length to total body length in M. malcolmsonii.

Figure 1.2: Allometric relation between RTL and CL in M.malcolmsonii

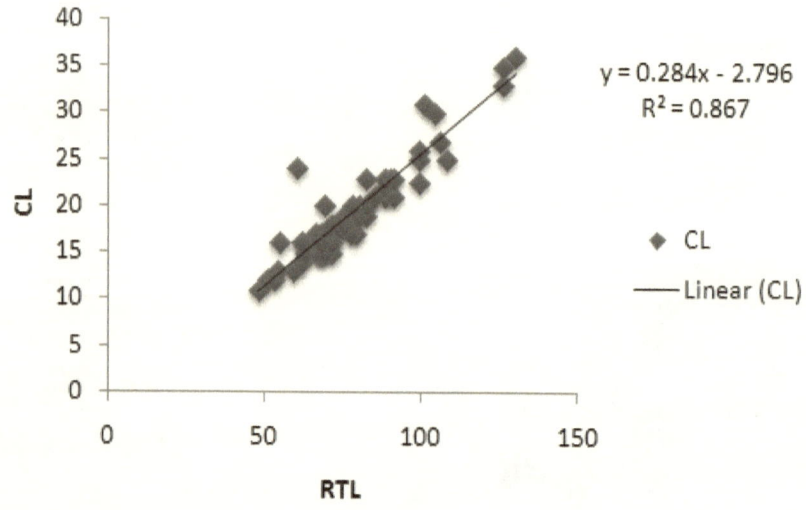

RTL vs Thoracic Length (ThL):

The regression is a good fit and on an average the ThL increased by 0.454 cm whenever RTL increases by one cm. In this model the intercept is negative (-0.908) which means that there is a minimum value of RTL at which the ThL will be positive (Figure 1.3). The correlation between RTL and ThL is 0.900. Slope value of 0.454 between ThL to RTL not only indicates positive allometry but also represents that higher growth of ThL may be due to longer rostrum.

Figure 1.3: Allometric relation between RTL and ThL in *M.malcolmsonii*

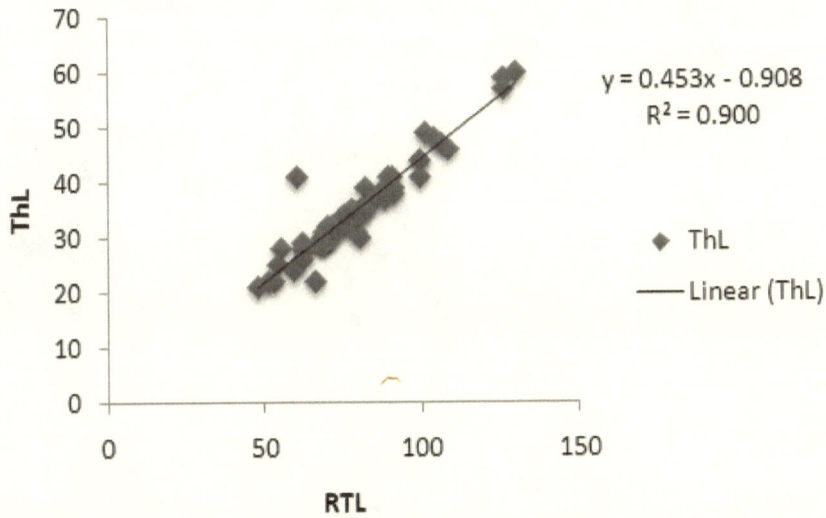

RTL vs Chelate leg Length (ChL):

The regression is a good fit and on an average the ChL increased by 1.047 cm whenever RTL increases by one cm. In this model the intercept is negative (-34.338) which means that there is a minimum value of RTL at which the ChL will be positive (Figure 1.4). The correlation between RTL and ChL is 0.849. Further the positive allometric relation between RTL and ChL clearly indicates that growth of the chelate leg is very prominent as a function of age in *M. malcolmsonii* which is highly advantageous in trapping big food particles and also for selective feeding. This is in agreement with the reports of George (1972).

Figure 1.4: Allometric relation between RTL and ChL in M.malcolmsonii

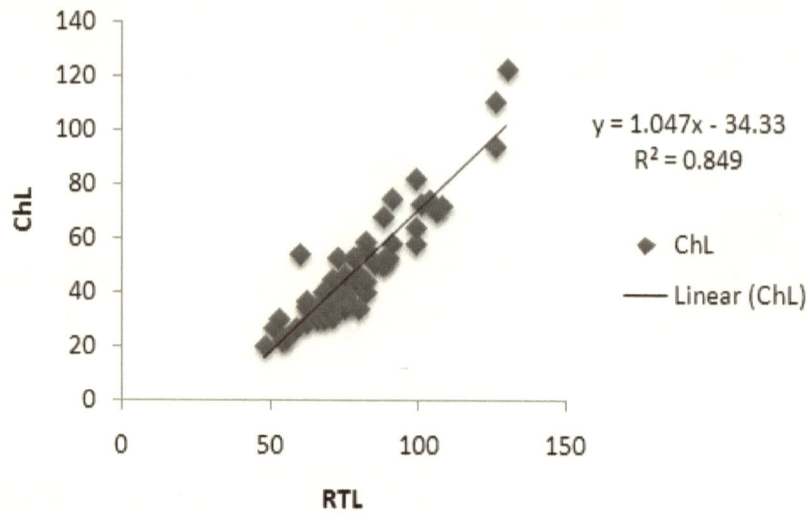

Rostral Length (RL) vs Carapace Length (CL):

The regression is a good fit and on an average the RL increased by 1508 cm whenever CL increases by one cm. The correlation between RL and CL is 0.760. The relationship between RL and CL was also found to be positively allometric in *M. malcolmsonii* (Figure 1.5). Lower correlation co-efficient (0.760) of RL to CL indicate less dependency of rostral length on the length of carapace. Due to the importance of rostral size in ethological efficiency of an organism, such as predation and burrowing activity, it is essential to note that the prediction of any relationship between RL and CL is enigmatic and that the cognizance of growth of rostrum to carapace is some what riddle to explain.

Figure 1.5: Allometric relation between CL and RL in *M.malcolmsonii*

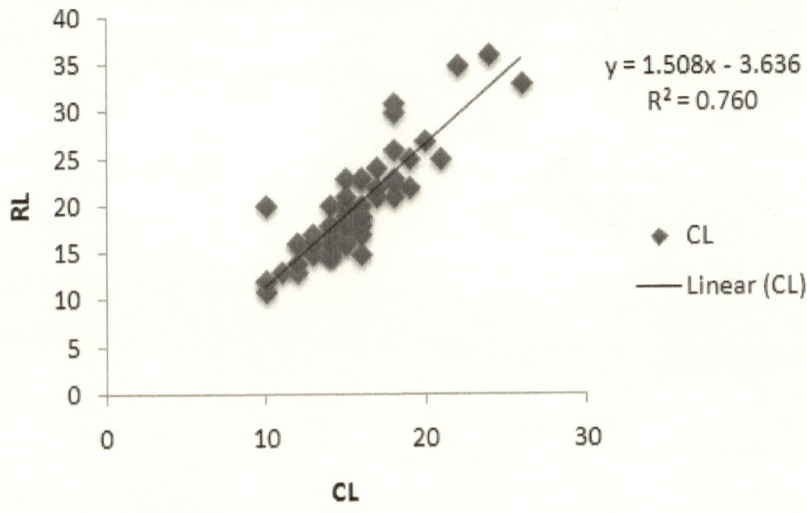

Biomass vs ThL:

The regression is a good fit and on an average the ThL increased by 1.524 cm whenever biomass increases by one g. The correlation between biomass and ThL is 0.884 (Figure 1.6). On the other hand a positive relation found to exist between whole animal biomass and ThL further explains the contribution of length of cephalothoracic region to the whole body biomass along with total body length. Thus the regression analysis carried out in the present study confirms species specific growth pattern in M. *malcolmsonii*.

The fact that all morphological characters studied in the present study account for positive allometry in M. *malcolmsonii* suggests that growth in this species is rapidly attained over a restricted series of moults. In general, *M. malcolmsonii* is usually larger and the carapacial growth is positively related to RTL.

Figure 1.6: Allometric relation between Biomass and ThL in *M.malcolmsonii*

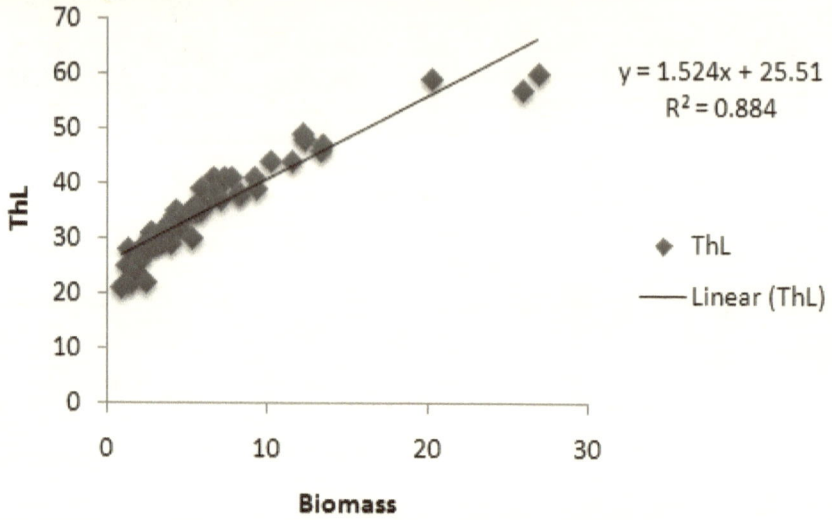

The original mass of many crustacean species could be predicted from the size of their body fragments (Richardson et al. 2000). It is a recognised aspect in order to determine growth rate and/or production of aquatic microinvertebrates as well as to understand life histories and trophic relations between functional and feeding groups (Benke, 1996). For this to happen higher number of samples must be taken to overcome spatial heterogeneity, when sampling a river, precludes the attempt to separate, identify and weigh animals while still alive (Cressa, 1986).

Measuring body structure of a single insect/prawn is less time consuming than weighing individually all the organisms collected. Since fundamental relations exist between linear body measurement (body length, head length, head width etc.) and biomass (Gould, 1966; Peters, 1983), they have been used in order to obtain indirect biomass estimates. Wenzel et al. (1990) indicated that a difference of 20% could be expected between actual and theoretical biomass.

Investigations carried out to explain similar type of relationships in other species reveal that otolith size is correlated with the size of the fish (Farrell & Campana, 1996). The relationship is a prerequisite for back calculation of the size of a fish at a prior age (Campana & Neilson, 1985; Jones, 1992), but recent studies have demonstrated that otolith size reflects fish size only loosely.

The conclusions, ofcourse, are based on the assumption that the morphometric data obtained from *M. malcolmsonii* reflected the average rate of change for different sized individuals, which inturn is presumably affected by similar average environmental conditions of its habitat. As pointed out by Kaufmann (1981) cross sectional data of this kind is easy to obtain but longitudinal data (i.e. consecutive moulting series as obtained either from nature or in the laboratory) are certainly more accurate. However such data is subject to limitations and may be variable. Owing to changes in environmental conditions in the field to which the prawns respond during the life time, or more importantly, because the growth series obtained under more or less constant experimental conditions may accurately be less realistic than those taking place in a typically fluctuating natural environment.

Thus the methodology adopted in this study allows at least a first approximation of growth for this less known prawn. The longitudinal data will provide further information confirming or denying this supposition.

Study of allometric relationship between various morphological features in *M. malclomsonii* provided some interesting trends. The equations could be useful to predict growth phenomenon and also of its subsequent maturity in the species. It is well known that a plot of 'a' versus 'b' for all known length weight relationships of a species results in a leniar relationship and that this relationship can be used to identify outliers (Froese and Pauly, 1998). As suggested by Ramamurthy and Manikkaraja (1978) this aspect of study is more significant where the size composition of the cultured prawns affect the dynamics of their resources.

Chapter 2: Characterization of moult cycle

Moult cycle stages in M. malcolmsonii (biomass 2.67±0.09g) were characterized using cuticle hardness or rigidity and changes in the developing setae of uropod as the main criteria. The moult cycle in M.malcolmsonii consists of four major morphologically distinguishable periods viz. postmoult, intermoult, premoult and ecdysis. While the external characteristic features such as shell hardness and decalcification of the carapace are evident at intermoult and postmoult stages, setal development provided accurate and convenient evidence for the premoult sub-stages. M. malcolmsonii was found to be one of the best suited decapod species for setogenesis of moult staging system and thus different subdivisions of the moult cycle stages could be identified using changes in the uropod morphology such as the retraction of the epidermal layer from the old cuticle as well as loosening of the old cuticle before ecdysis along with changes involved in setogenesis.

Post moult

Stage A1:

Postmoult stage A1 refers to the stage that manifests immediately after ecdysis. Generally prawns of this stage were inactive, did not consume food for about 30min and were found to regain activity thereafter. No internal cones in setae were found (Figure 2.1A). During the entire A1 stage which lasted for 4-5hr the exoskeleton was very soft and pliable over the whole carapace. Any sort of stress due to handling was found to cause mortality at this stage. In freshly moulted condition the animal increased in size due to oral intake of water during the active phase of ecdysis as well as diffusion through cuticle.

Stage A2:

The carapace is found to be still pliable and transparent. The epidermis of the uropod closely applied to the cuticle with the setal lumen having a wide diameter and is prominent with granular matrix. The granular protoplasmic matrix is continuous throughout the setae filling the new setal articulation and distal end of the setae. Exoskeleton started to harden. This stage lasts for 10-12 hr.

Figure 2.1A: Characterization of Postmoult stage A (X50)

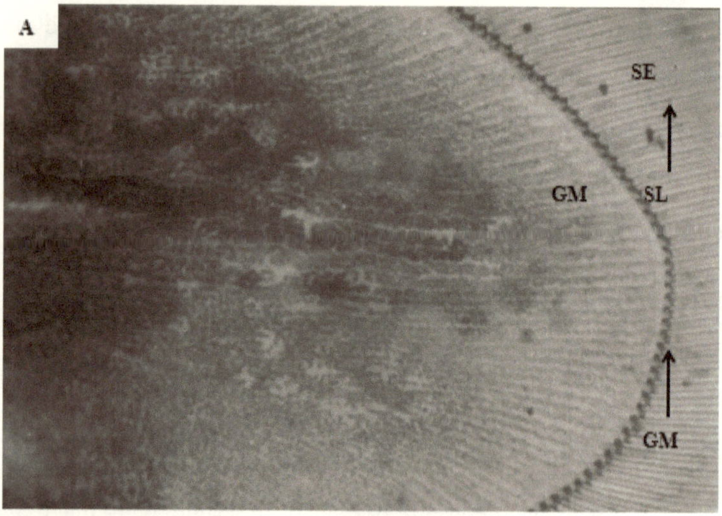

SE: Setae; GM: Granular matrix; SL: Setal lumen.

Stage B:

This stage is characterized by hardened, yet easily depressible carapace. At this stage, the dorso lateral sides of the carapace continue to harden but it could be depressed with finger pressure. Setal cones are not formed in the uropod. This stage is marked by the appearance of well-developed setal articulation and the beginning of cuticular node development. The setal protoplasmic matrix is found concentrated within the setal lumen so as to fill only the proximal half of the setae (Figure 2.1B). This stage lasted for 24 hr.

Figure 2.1B: Characterization of Postmoult stage B (X100)

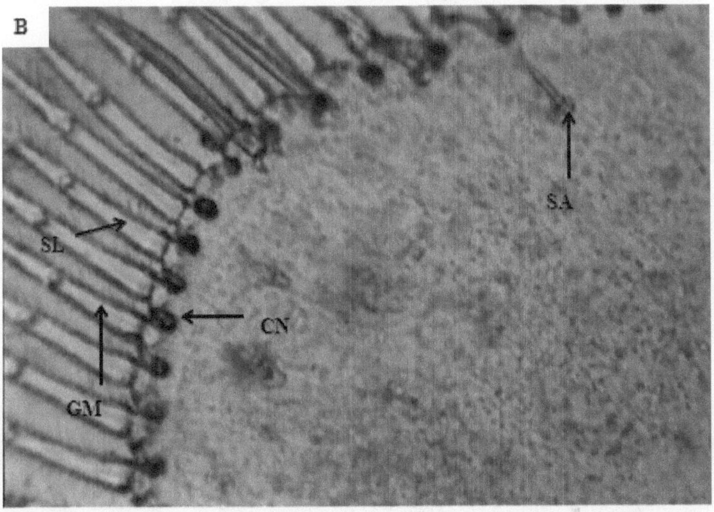

GM: Granular matrix; SL: Setal lumen; CN: Cuticular node; SA: Setal articulation.

Intermoult C:

In this stage the cephalothorax and the appendages appear uniformly hard and the prawn started to feed actively. The exoskeleton has become hard and calcified progressively and hence identification of further substages of this stage becomes difficult based on the setal development. The characteristic feature of intermoult stage is that the setogenic events in the uropod have almost been completed. Prominently visible internal cones are present inside the setae (Figure 2.1C). Tissue filling occurred at the base of the setae. Rostrum is firm.

Epidermis is not retracted from the cuticle between the setae. This stage lasts for 3-4 days.

Figure 2.1C: Characterization of Intermoult stage C (X100)

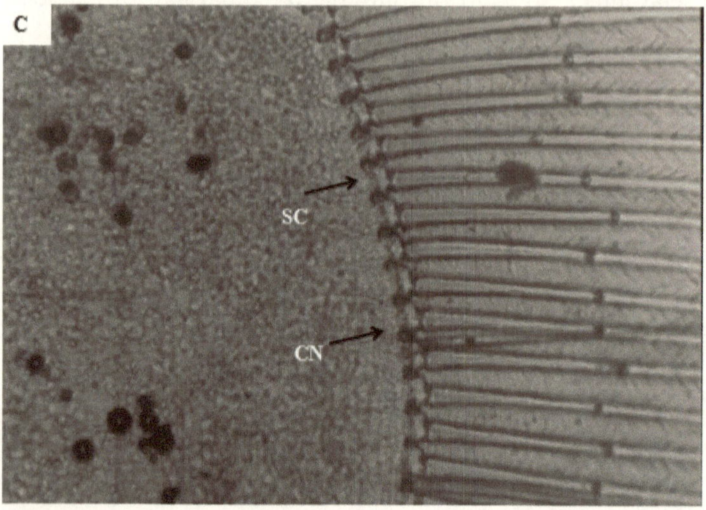

SC: Setal cone; CN: Cuticular node

Premoult:

The premoult stage is the preparatory stage for the ensuin moult. Premoult is an active stage of the moult cycle encompassing old cuticle resorption and new cuticle synthesis during which calcium resorption from the shell occurs and new cuticle is secreted (Vijayan et al. 1997). The animal stops feeding at some point of time during this stage. The first morphologically distinguishable evidence of premoult stage (Figure 2.1D, E.F, G, H) started with apolysis, the retraction of epidermis from the cuticle, creating a moulting space for the formation of new cuticle. The epidermal retraction followed by the secretion of new cuticle could be easily seen in the uropod. Hence uropodal examination during premoult stage (D) has yielded several substages such as D1, D2, D3 and D4 which lasted for 14.45 days. Furthermore, there is a sequential change in the development of new setae from the presumptive setal grooves. Based on the above characteristic features the premoult stage is divided into the following substages.

Stage Do:

In this stage, the rigidity of the exoskeleton is same as stage C. The first indication of this stage is the retraction of epidermis from the cuticle at the base of the setae (visible initiation of Apolysis in the uropod) as well as pigment retraction. Apolysis is evidenced by the appearance of crescent shaped gap just below the tip of the uropod cuticle (Figure 2.1D). The retraction of epidermis from the cuticle creates moulting space for the formation of new cuticle. Depending upon the extent of retraction a translucent "epidermal zone" bounded by a distinct epidermal line appears. This stage lasts for 2- 3 days.

Figure 2.1D: Characterization of Premoult stage D0 (X100)

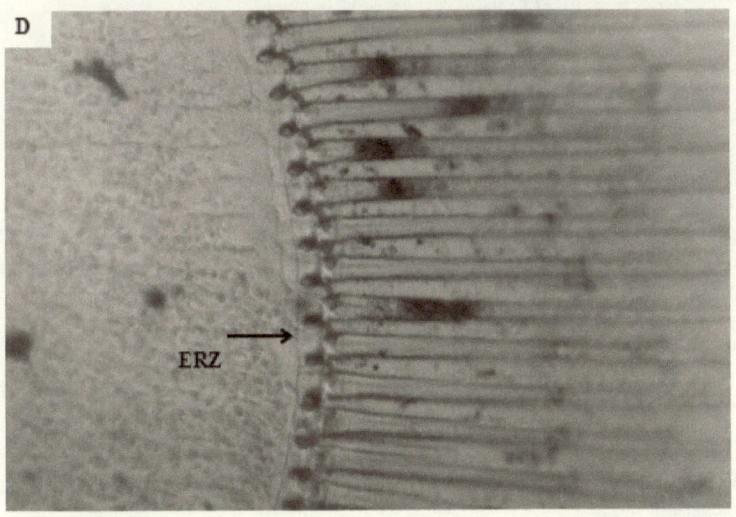

ERZ: Epidermal retraction zone

Stage D1:

The most characteristic feature of this stage is the appearance of an invagination at the site of each future setae (Figure 2.1E) signaling the new setal development within the setal grooves of the uropodal tissue. In this stage the cuticle is hard but gradually thinned and becomes brittle. All the appendages are rigid. When the setal groove reaches the periphery of the epidermis, the new juvenile setae would be completely protruded out into the retracted zone. The resorption of the old cuticle also started simultaneously. The animal stops to take food at this stage. This stage lasts for 2.1 days.

Figure 2.1E: Characterization of Prmoult stage D1 (X100)

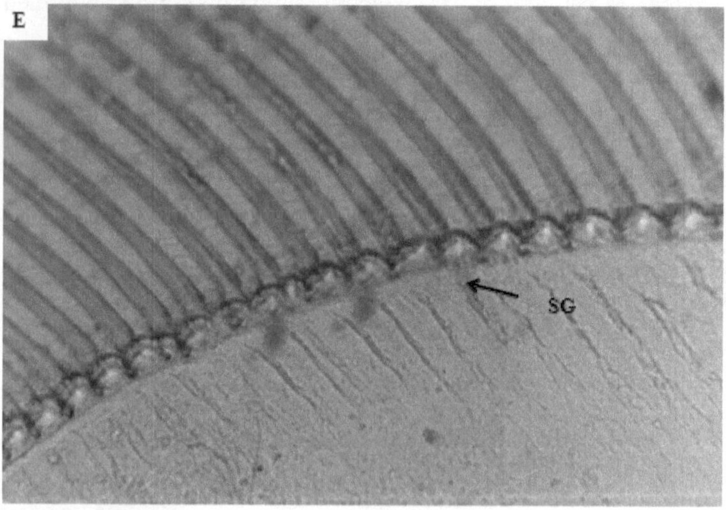

SG: Setal groove

Stage D2:

During this stage new setae are clearly seen and easily defined. The epidermis is regressed even further from the cuticle (Figure 2.1F) and continues to regress slowly until ecdysis as reported by Freeman and Bartell (1975). There is a marked change in the appearance of carapace, as the exoskeleton becomes more brittle. The new cuticle development occurred very rapidly at this point. The setal invaginations with setae are visibly thickened and clearly observed. This stage lasts for 3.1 days.

Figure 2.1F: Characterization of Premoult stage D2 (X100)

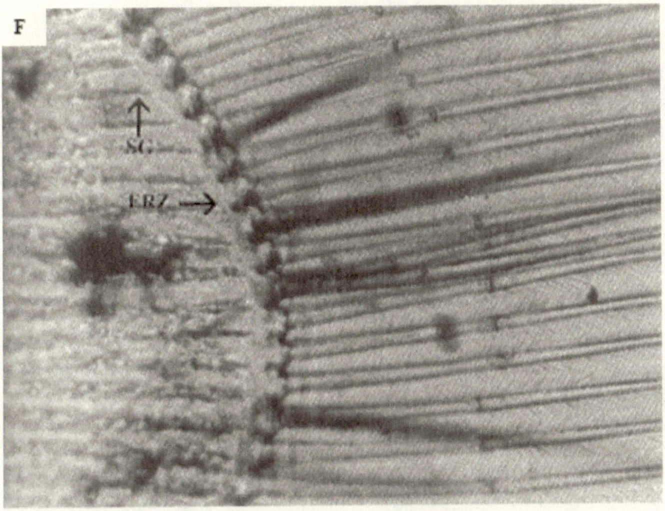

SG:Setal groove; ERZ: Epidermal retraction zone

Stage D3:

In this stage the new setae have extruded almost completely into the transparent retracted zone and setal articulation could be seen more prominently in the new cuticle. Bristles are evident on seta' shafts (Figure 2.1G). The cellular matrix is yellowish in colour. This stage lasts for 3.6 days.

Figure 2.1G: Characterization of Premoult stage D3 (X100)

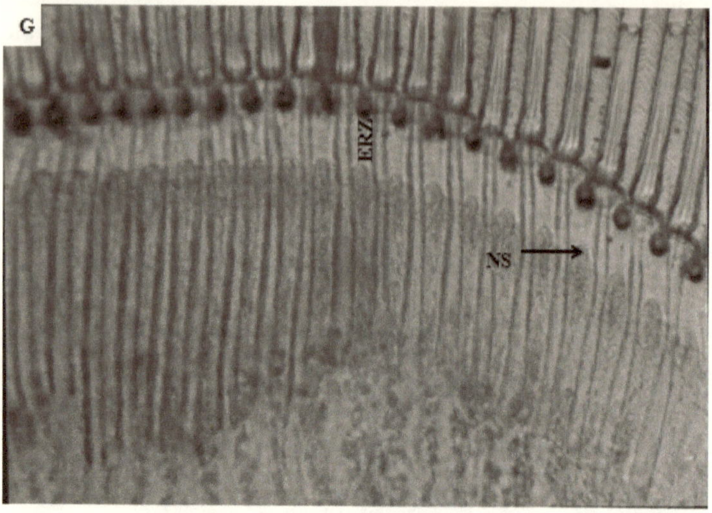

ERZ: Epidermal retraction zone; NS: New setae

Stage D4:

During this stage the old cuticle and setae could physically be separated from the new uropod. The newly forming cuticle has thick setae at the base of the uropod (Figure 2.1H). Disappearance of epidermal retraction zone is seen during this stage due to the movement of new setae to the periphery of uropod. Furthermore, the prawn is inactive during this stage and water absorption through new cuticle begins resulting in swollen body cavity. At the end of this stage prawns would be ready to shed off the exoskeleton. This stage lasts for 3.4 days.

Figure 2.1H: Characterization of Premoult stage D4 (X100)

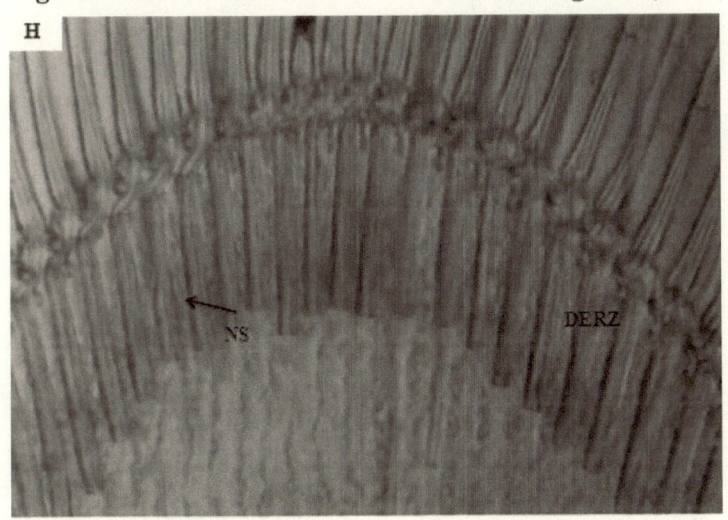

NS: New setae; DERZ: Disappearance of epidermal retraction zone

Ecdysis:

Due to endocuticular resorption of moulting fluid the exoskeleton becomes thin and friable during this stage. As a result prawns emerge through the ecdysial sutures formed in the intersegmental membrane connecting the cephalothorax and abdomen of the old exoskeleton.

After ecdysis the exoskeleton of freshly moulted prawn becomes soft, pliable and thin and is stretched to increase the body volume. As a result the size of the animal after ecdysis was found to be much larger than before. Finally the prawn comes out of the old cuticle and everts the setae of the new exoskeleton (Figure 2.1I), the old exoskeleton or exuvia was intact along with all appendages (Figure 2.1J). This stage last for 3-5 min and if it is prolonged it results in the death of the animal.

Figure 2.1I: Characterization of Ecdysis stage (X100)

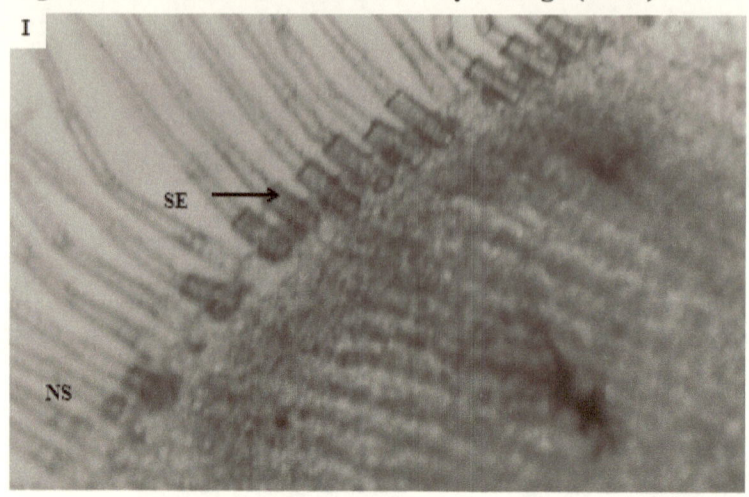

SE:Setae; NS: New setae

Figure 2.1J: Freshly moulted (Top; X2) along with exuvia of individual (Bottom; X2)

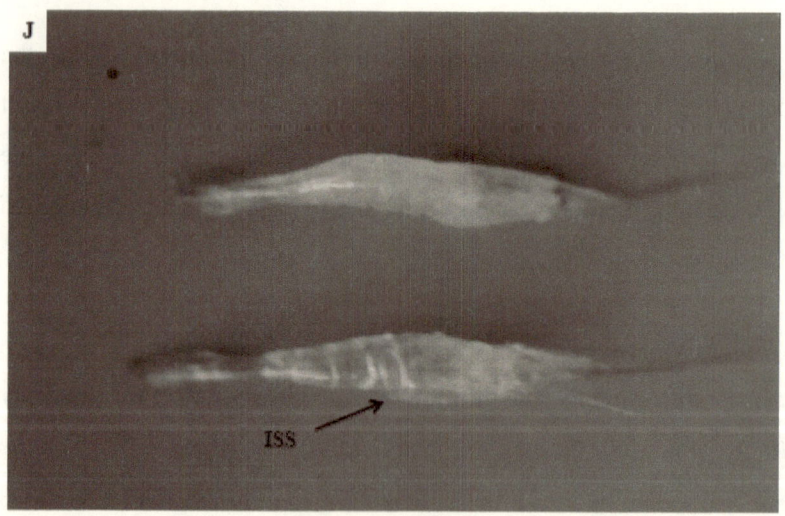

ISS: Inter segmental suture

Effect of eyestalk ablation on moult cycle:

Eyestalk ablation was performed in order to understand its effect on moult cycle of M. malcolmsonii. Changes in moult cycle duration of control and ablated M. malcolnisonii of four different size groups are presented in the figure 2.2a. Moult cycle duration in both control and ablated prawns increased significantly with increase in biomass. Smaller prawns moult faster with short moult cycle duration and larger prawns moulting slowly with longer moult cycle duration, is in confirmation with the general moulting pattern among the prawns (Dail et al. 1990; Vijayan et al. 1997).

Figure 2.2a: Total moult duration in control and eyestalk ablated *M. malcolmsonii* of four different size groups

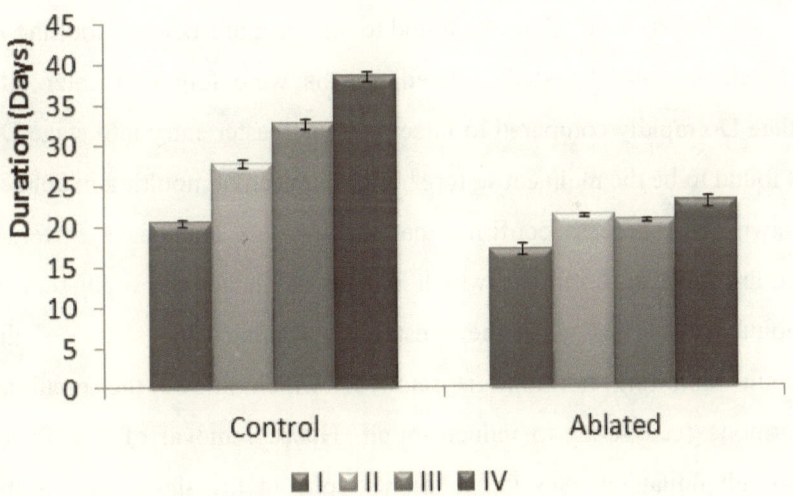

In each size group duration of the moult cycle was found to be significantly lower in ablated prawns compared to respective controls (Figure 2.2a). However percent change in moult cycle duration of ablated prawns increased with increase in size from group I to IV (Figure 2.2b).

Figure 2.2b: percent change in moult cycle duration of ablated prawns over controls in four different size groups

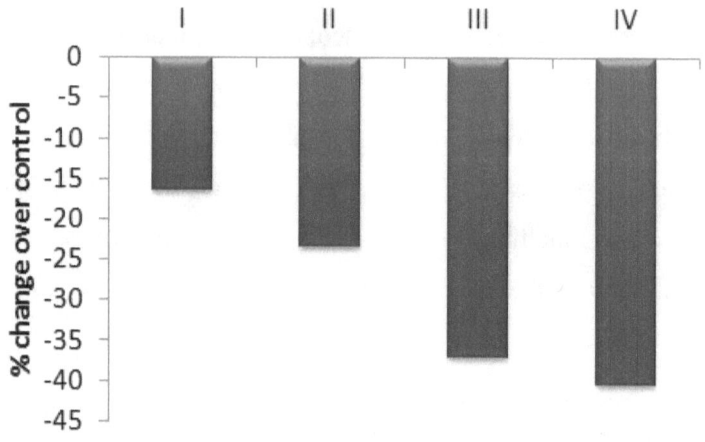

Eyestalk ablation was found to initiate a precocious moulting in M.malcolmsonii. Eyestalk ablated prawns were found to enter into stage Do rapidly compared to intact prawns. Faster entry into stage D0 is found to be the main cause for shorter duration of moulting in ablated prawn. This process confirms that the eyestalk contains a factor, a moult-inhibiting hormone, which is responsible for the regulation of moult cycle. Thus when the eyestalks are ablated the source of the moult inhibiting hormone is removed which allows the moulting hormone (ecdysone) to induce moult. Hence removal of this factor through unilateral eyestalk ablation results in the shortening of the intermoult and premoult stages (Figure 2.3). Many investigators in several crustaceans found that eyestalk ablation resulted in precocious moulting and shortening of the moult cycle duration (Chakravarthy, 1992; Chang, 1997; Sagi et al. 1997; Radiant et al. 2000).

Figure 2.3: Duration of moult cycle stages in normal and eyestalk ablated *M. malcolmsonii*

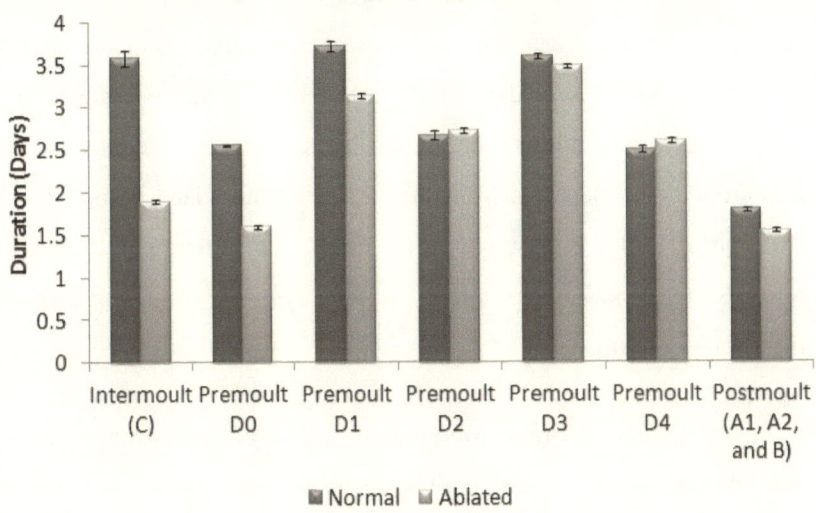

5-Hydroxytryptamine (5-HT) present in the central nervous system of crustaceans (Fingerman et al. 1994) has been proposed as a transmitter or modulator, mediating a variety of physiological functions in crustaceans (Saenz et al. 1997). Administration of 5-HT initiated synthesis and release of moult inhibiting hormone (MIH) and thereby inhibiting moulting (Mattson and Spaziani, 1985; Sarojini et al. 1994). Removal of 5-HT through eyestalk ablation obviously eliminates 5-HT action on eyestalk neuro-secretory cells leading to secretion of ecdysteroids which inturn initiate moulting.

Moulting in crustacea is positively regulated by ecdysteroids secreted by the Y-organ, an epidermal derivative equivalent to the prothoracic glands of insects (Gabe, 1953; Spaziani et al. 1989). The synthesis and secretion of ecdysone by Y-organ is under the inhibitory control of a neurohormone originating in the X-organ sinus gland of the eyestalk (Soumoff and 0' Connor, 1982; Webster, 1998). Eyestalk ablation removes the source of inhibition not only of the Y-organ but also of the MO. Recent studies have also indicated that Y-organ activity is further under the influence of a stimulatory factor, methyl farnesoate (MF) secreted from the mandibular organ (Chang, 1993). Interestingly, mandibular organ itself is negatively controlled by inhibitory neuro-peptides from the x-organ sinus gland complex (Liu and Laufer, 1996; Wainwright et al. 1996). Hence removal of eyestalk releases the inhibitory action of the eyestalk neuropeptides on mandibular organ.

Eyestalk ablation was found to induce hypertrophy of the mandibular organ causing ultrastructural changes in the nuclei, mitochondria and endoplasmic reticulum (Hinsch, 1977; Taketomi and Kawano, 1985). These changes have been interpreted as signs of increased synthetic activity of the mandibular organ. Measurements of MF synthesis and MF titre following eyestalk ablation in L. emarginata have confirmed this hypothesis (Laufer et al. 1986, 1987; Tsukimura and Borst, 1992; Homola, 1997).

Earlier reports indicating a significant decrease in the moult interval in white shrimp, Penaeus setiferus that were implanted with mandibular organs from the blue crab, Callinectes sapidus (Yudin et al. 1980) and induction of premoult in Caridina denticulata by the injection of mandibular organ homogenates from P. clarkii (Taketomi et al. 1989) also provided evidence regarding the role of MF in the regulation of moult cycle among crustaceans. Though direct measurements of 5-HT or MIH have not been made in the present study, the effect of eyestalk ablation on the duration of moult cycle stages provide enough evidence regarding their role.

Chapter 3: Bioenergetics

A bioenergetic model could be used to predict growth of an organism on the basis of known food consumption or to derive food consumption estimates based on observed growth (Hewett and Johnson, 1992). The food consumed by an animal and its relationship to growth are of particular importance, since these are crucial for understanding and modeling of aquatic ecosystems (Steele, 1974). Food consumption may vary over daily or seasonal time intervals, in response to prey availability, weather and many other intrinsic factors such as body weight, reproductive status, hunger level etc. The main energy input or income to the budget of an organism is the food consumed. The catabolism of food is organised within the animal to conserve free energy for use in anabolic and life sustaining processes. Majority of field studies on food niches or feeding regimes have ignored temporal differences in feeding and intraspecific weight (size or age) class differences in utilization. Food intake can be measured from the quantities of food contained within the stomach (Elliott and Persson, 1978; Man, 1978; Windell, 1978) or by estimating the rate of gastric evacuation (Eggers, 1977; Elliott and Persson, 1978; Olson and Mullen, 1986; Boisclair and Legget, 1988; Heroux and Magnan, 1996) along with measuring a change in the stomach contents. Alternatively, in laboratory studies, food ingestion is estimated by providing measured quantities of food with known energy content. This method has been followed in the present investigation.

Changes in the amount of energy consumed (C) in control and ablated *M. malcolmsonii* of four different size groups are presented in figure 3.1a respectively. Amount of energy consumed by both control and ablated prawns increased significantly (3.1b) while the rate of consumption decreased with increase in biomass (Figure 3.1c).

Figure 3.1a: Energy consumption, absorption, assimilation and energy available for growth (J / day) in control and eyestalk ablated *M.malcolmsonii*

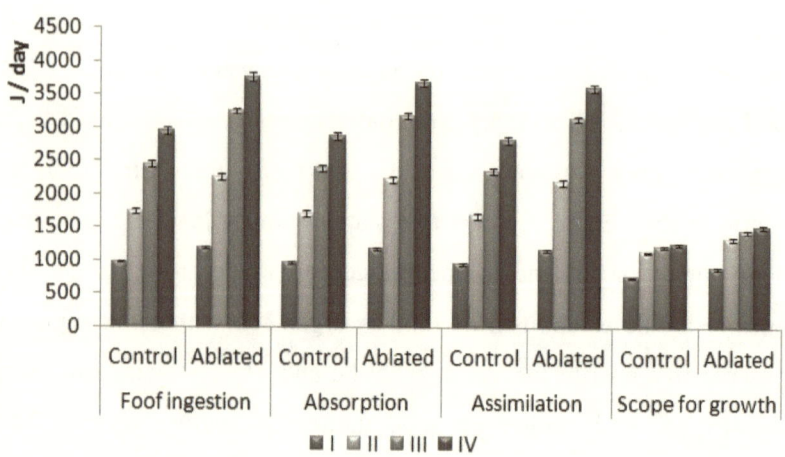

This has been the regular trend noticed in different organisms in several of the earlier studies. Measurements among fish showed increased food consumption with increase in biomass in the chinook salmon, Oncorhynchus tshawytscha (Walbaum) (Davis and Warren, 1968), coho salmon, *O. kisutch* (Edsall et al. 1974), large mouth bass, *M. salmoides* (Rice and Cochran, 1984), northern pike, *E. lucius* (Diana, 1983) and cyprinid, *P. phoxinus* (Cui and Wootton, 1988). Griffiths (1991) observed that percentage energy intake declined with increase in body size in ant- lion larvae. Body weight dependence of maximum food consumed (C_{max}) in fish represented by power function was also found to vary depending upon ration and temperature (Brett, 1971; Elliott, 1976; Beauchamp et al. 1989). Fischer (1973) and

Elliott (1979) have used $C_{max} = a\, w^b$ to describe the relationship between maximum rate of food ingestion and biomass for several species including grass carp, *Ctenopharyngodon idellus* (Valenciennes) and brown trout, *S. trutta* (Linnaeus). Increased energy acquisition with increase in biomass has also been noticed in shellfish such as the freshwater prawn, *M. rosenbergii* (Sierra and Diaz, 1999) and the gastropods, *Nucula nitidosa* (Winckworth) and *Nucula turgida* (Leckenby and Marshall) (Rachor, 1976; Davis and Wilson, 1985; Wilson, 1988).

Figure 3.1b: Percent change in energy consumption, absorption, assimilation and energy available for growth of ablated prawns over controls.

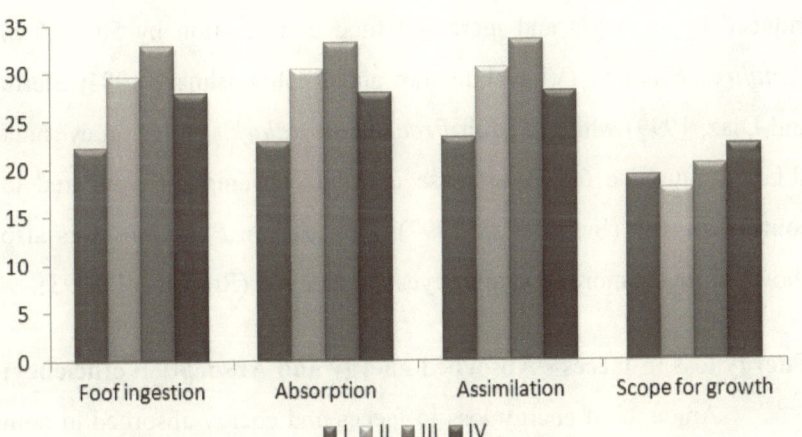

Measurement of rate of consumption (RC) provided a clear evidence that consumption levels per unit body weight decrease gradually as the size of the prawn increases which could be due to lack of increase in gut capacity in proportion to the increase in body size. Ablation did not seem to affect the above relation. Changes in the percent increase in energy consumed between two successive size groups (Figure3.1b) further demonstrated the magnitude of decrease in increased food consumption with increase in biomass. Control prawns showed 78% increase in energy consumption between size groups I and II, 41% increase between II and III and 20% increase between III and IV. Eyestalk ablated group on the other hand showed 88% increase in energy consumption between size groups I and II, 43% increase between II and III and 17% increase between III and IV.

In each size group the amount of energy consumed is significantly higher in eyestalk ablated prawns compared to the respective controls (Figure 3.1b). However the percent change in energy consumption of eyestalk ablated prawns over the respective controls increased with increase in biomass upto size group III, but decreased significantly at size group IV (Figure 3.1c). Eyestalk ablation induced hyperphagia and increased food consumption by 50-70% in *Panulirus homarus* (Vijayakumaran and Radhakrishnan, 1984; Sierra and Diaz, 1999) while in adult *Procambarus clarkii* unilateral eyestalk ablation caused a 6.8 % increase in food consumption compared to control crayfish (Sierra et al. 1997). Pink shrimp *P. notialis* was also shown to ingest more food after eyestalk ablation (Rosas et al. 1993).

Energy loss in faeces - Absorbed energy and Absorption efficiency:

Amounts of energy loss in faeces and energy absorbed in both control and eyestalk ablated *M. malcolmsonii* are presented in figure 3.1a and 3.2a respectively. A significant increase was noticed in energy loss in faeces and energy absorbed with increase in biomass both in control and eyestalk ablated prawns. However the rate of increase in energy loss in faeces and the rate of increase in energy absorbed between two successive stages decreased significantly with increase in biomass in both the groups.

Although no significant effect of biomass on faeces production has been found in the cyprinid, *P. phoxinus* (Cui and Wootton, 1988) a linear relationship between faeces production and food composition was found in several other fish (Gerking, 1955; Cui and Wootton, 1988). Faecal material in animals may arise in four ways (Brafield and Llewellyn, 1982). First, some ingested food may not be digested because of absence of required enzymes. Second, the retention time of food in the alimentary canal may be too short. Third, some of the products of digestion may not be assimilated and fourth, molecules capable of assimilation may not actually be assimilated. Thus faeces are one of the enzyme-rich waste products eliminated from the animal which is a mixture of undigested food components and unabsorbed residues. For an energy budget the measurement of faeces is essential because it has separate origins and distinctive chemical composition and thus possesses different energy contents.

Figure 3.2: Absorption efficiency and assimilation efficiency (%) in control and ablated *M.malcolmsonii*

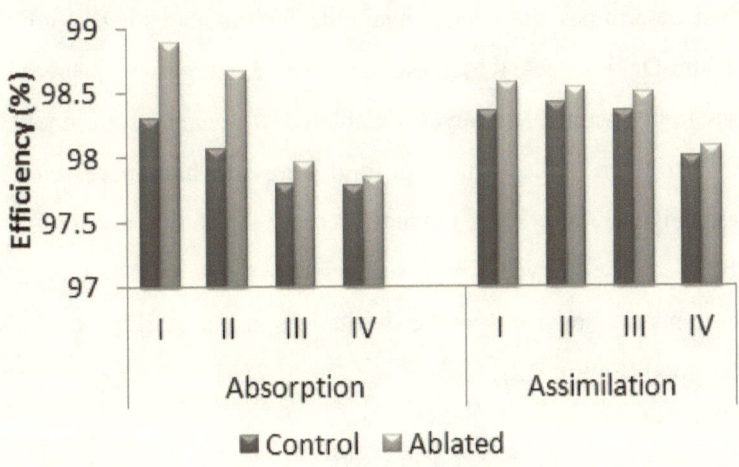

Further the quantity of faeces depends not only on the nature of ingested food but also on the extent of digestion and absorption. Hence marked increase in energy loss in faeces in *M. malcolmsonii* with increase in biomass in control and eyestalk ablated groups (Figure 3.2a) indicate that larger individuals lose greater amounts of energy in egestion, may be due to their inability to elevate digestion and /or absorption efficiency on par with the ability to consume large amounts of food. While in control prawns the rate of increase in faeces between size group I and II was 103%, it was 61% between II and III and 24 % between III and IV. In ablated prawns it was 130% between size group I and II, 115% between II and III and 23% between III and IV. This decreasing trend clearly indicates that *M. malcolmsonii* exhibit a compensatory mechanism in digestion and absorption with increase in biomass, perhaps to avoid excess loss of energy through faeces especially with progress in life history stages.

Digestion and absorption mainly occur in the intestine although crop is also involved in predatory species for this purpose. Energy absorbed determines the energy available for metabolism and growth (Smith and Davies, 1995). Increase in absorbed energy with increase in biomass in both control and eyestalk ablated *M. malcolmsonii* could be directly attributed to the quantity of food ingested which increased with biomass (Figure 3.1a). But a gradual decrease in the rate of increase in absorbed energy between two successive stages with increase in biomass indicate a size specific limitation in the ability to absorb ingested food (Figure 3.2).

Energy loss in faeces decreased in eyestalk ablated prawns of size group I and II and increased in III and IV compared to the respective controls (Figure 3.2b). Although eyestalk ablated prawns of all size groups showed significantly higher levels of absorbed energy compared to the respective controls (Figure 3.1a), the percent change in energy absorbed over controls increased from size group I to III but decreased at size group IV (Figure 3.1b). These results further indicate that although eyestalk ablation cause hyperphagia in prawns of all size groups, larger prawns might not be able to maximise digestion and absorption abilities along with increased ability of ingestion compared to prawns of earlier life stages.

The potential amount of energy loss in excretion and energy available for metabolism and growth depend upon the proportion of food energy lost in egestion. Elliott (1976) found that 25-30% of food energy was lost in egestion in *S. trutta* feeding on *Gammarus* sps. while 6.5% was lost in *minnows*, feeding on white worms (Cui and Wootton, 1988) and 25.6% was lost in the fish, *R. rutilus* feeding on meal worm (Hofer et al. 1985). Based on the present data (Figure 3.2a) it is clear that depending upon the body size, *M. malcolmsonii* looses 1.7 to 2.3% of ingested energy in faeces resulting in 97.8 - 98.3 % of ingested energy available for other energy demanding processes. Lower proportions of energy loss in faeces in eyestalk ablated *M. malcolmsonii* compared to the respective controls may be another adaptation towards meeting the enhanced energy demands.

Absorption efficiency, on the other hand, decreased with increase in biomass in both control and eyestalk ablated prawns with the level of absorption efficiency being higher in eyestalk ablated prawns at each size. Decrease in absorption efficiency with increase in biomass in both the groups could well be the reason for increased loss of faeces with increase in biomass (Figure 3.2b). Although Elliott (1976) found no effect of biomass on absorption efficiency in *S. trutta*, Kelso (1972) noticed decreased absorption efficiency in *S. vitreum* with increase in biomass. In contrast, Smith (1973) in *Histrio histrio* (L.), Allen and Wootton (1983) in *G. aculeatus* and From and Rasmussen (1984) in *S.gairdneri* observed an increase in absorption efficiency with increase in biomass. Increased absorption efficiency has been noticed earlier in both male and female *M. nobilli* upon eyestalk ablation (Sindhu Kumari and Pandian, 1987). Similarly unilateral eyestalk ablation in the juvenile blue crab, *Portunus pelagicus* (L) caused a significant increase in absorption efficiency and dry meat yield (1% of dry weight) (Germano, 1994). Significantly higher levels of absorbed energy and absorption efficiency in eyestalk ablated *M. malcolmsonii*, show that higher energy needs related to tissue repair / regeneration caused due to ablation and compensatory mechanisms involved due to the loss of neurosecretory centres are met not only through increased food ingestion but also by increased absorption efficiency.

Energy loss in excretion - Assimilated energy and Assimilation efficiency:

Many of the simple products of digestion which are absorbed by the gut are broken down sooner or later. The waste products of carbohydrate and fat metabolism are carbondioxide and water, but in the case of protein some nitrogenous wastes are produced as well, which contain some energy. The commonest amongst such waste products are ammonia, urea and uric acid derived either from absorbed food or from body tissues mainly composed of carbon, hydrogen, oxygen and nitrogen. Nitrogen containing compounds are especially important for animals because they are essential for making proteins and nucleic acids, necessary for maintenance and growth. Aquatic organisms are generally ammonotelic. Ammonia is the major excretory product in freshwater invertebrates since it is readily eliminated from the body (Zebe et al. 1986; Pandian, 1987; Randall and Wright, 1987; Regnault, 1987). Nitrogen excretion has two components: endogenous excretion and exogenous excretion. Endogenous excretion is defined as the minimum value of nitrogen excretion and is estimated after complete starvation or fed without proteins (Savitz, 1971). It also corresponds to the degradation of body proteins after the utilization of reserves. Exogenous nitrogen excretion refers to the excretion of nitrogen from the food source and is mainly influenced by food consumption (Elliott and Persson, 1978; Savitz et al. 1977; From and Rasmussen, 1984).

Data related to the amount of energy loss in excretion and assimilated energy in both control and eyestalk ablated *M. malcolmsonii* are presented in figure 3.3a and 3.1a respectively. A significant increase was noticed in energy loss in excretion and also in assimilated energy with increase in biomass both in control and eyestalk ablated prawns. However the rate of increase in energy loss in excretion and increase in assimilated energy between two successive stages decreased with increase in biomass in both the groups (Figure 3.2).

Davenport et al. (1990) reported that ammonia output in the captive Atlantic halibut, *H. hippoglossus* and lemon sole *M. Kitt* approximately doubled 24hr after a meal. An increase in ammonia excretion-indicating increased energy costs-following food ingestion has been observed earlier in several species (Brett and Zala, 1975; Caulton, 1978; Zebe et al. 1986; Gonzaleg et al. 1990) although little effect of body weight on energy loss in excretion has been noticed in *S. trutta* (Elliott, 1976). Stressful conditions were also found to cause an increase in nitrogen loss in fish (Hunn, 1982) thus enhancing the loss of combustible matter.

Crustaceans are generally ammonotelic mainly converting the end product of nitrogenous catabolism to ammonia (Claybrook, 1983). In decapod crustaceans 60 - 70% nitrogen is mainly excreted as ammonia (Regnault, 1987). Environmental factors like temperature, salinity and ammonia have earlier been reported to affect ammonia excretion of crustaceans and were reviewed by Regnault (1987). Increase in energy lost in excretion with increase in biomass in control and eyestalk ablated *M.malcolmsonii* could be attributed to increased food ingestion and thus to the rate of protein catabolism. While in control *M. malcolmsonii* the increase in excretion was 71% between size group I and II, 46% between II and III and 45% between III and IV, in eyestalk ablated prawns it was 94% between size group I and II, 46% between II and III and 48% between III and IV. Gerking (1955) demonstrated a weight dependent relation for endogenous nitrogen excretion in bluegill sunfish following the well known relationship of a decrease in the rate of metabolism with increase in body size. In copepods also the excretion rate was found to increase allometrically with individual length (Nival et al. 1974; Ikeda, 1985) while the excretion rate per unit weight decreased with (Corner et al. 1965) or was independent of individual length (Kremer, 1977; Morand et al. 1987).

Similarly Nelson et al. (1977) found a positive correlation between ammonia production and biomass in *M. rosenbergii* although unit excretion showed a negative trend. *P. monodon* and *P. semisulcatus* (De Hann) also showed a decrease in the rate of ammonia excretion with increase in biomass (Wickins, 1985; Wajsbort et al. 1989). Further in *P.chinensis* (De Hann) ammonia - N excretion had a negative power relationship with in the range from 0.29 to 11.19g biomass indicating a positive correlation between total excretion and biomass, a general trend observed in several aquatic organisms (Gerking, 1955; Iwata, 1970; Chen et al. 1991) although fractional loss of excretion was found to be independent of weight (4 to 300 g range) in gammarids (Elliott, 1976).

Movement of energy within an animal depends on the efficiency with which it assimilates energy from the ingested food and the efficiency with which the assimilated energy is converted to production. The energy actually assimilated by the animal is the difference between the energy absorbed and energy lost in excretion. Petrusewicz and Macfadyen (1970) regarded assimilation as the energy channelled into respiration and production. Increase in assimilated energy (As) with increase in biomass in both control and eyestalk ablated *M. malcolmsonii* (Figure 3.1a) along with an increase in food ingestion indicate that the energy assimilated is primarily an effect of the amount of food ingested.

Eyestalk ablated prawns showed higher levels of energy loss in excretion at each size group over the controls (Figure 3.3b). Rate of ammonia excretion has been reported to be higher in eyestalkless crustaceans than in intact individuals (Bliss, 1953; Rahavaiah et al.

1980; Koshio et al. 1992b). Eyestalk ablation in the juvenile lobster, *H.americanus* lowered nitrogen excretion than found in intact lobsters (Koshio et al. 1992b). The energetic costs of ammonia excretion were found to be significantly lower in eyestalk ablated males and females compared to intact crayfish (Sierra et al. 1997). However eyestalk ablation was found to cause an increase in ammonia excretion in the freshwater prawn *M. lanchesteri* (Raman et al. 1981; Ananthakrishnan et al. 1981), the shrimp *P. indicus* (Kiron and Diwan, 1985), *P. monodon* (Nan et al. 1995) and *P. japonicus* (Bate) (Chen and Chia, 1996) indicating that unilateral eyestalk ablation significantly affects nitrogen metabolism. Regnault (1979) explained that removal of eyestalk neuro-endocrine system may elaborate the principle (s) responsible for the regulation of NH_3 excretion.

Increased food consumption (Figure 3.1a) and increased energy loss in ammonia excretion (Figure 3.3a) as a result of eyestalk ablation in *M. malcolmsonii* suggest increased catabolism and utilization of proteins due to higher energy demands. Greater percent elevation in energy loss in excretion in eyestalk ablated prawns of different size groups compared to the respective controls (Figure 3.3b) further provide evidence for the combined effect of eyestalk ablation and biomass. Despite higher loss of energy in faeces (Figure 3.3b), the amount of assimilated energy is higher in eyetalk ablated prawns over the control prawns and increased with increase in biomass at least upto size group III (Figure 3.1b). Each size group of both control and eyestalk ablated *M. malcolmsonii* was found to exhibit more than 95% assimilation efficiency indicating availability of greater proportion of ingested energy for metabolism (R) and growth (G). Rosas et al. (1993)

reported increased assimilated energy and higher scope for growth in the shrimp, *P. notialis* upon eyestalk ablation.

Figure 3.3a: Energy loss in faeces, excretion, routine metabolism and ASDA (J / day) in control and eyestalk ablated *M.malcolmsonii*

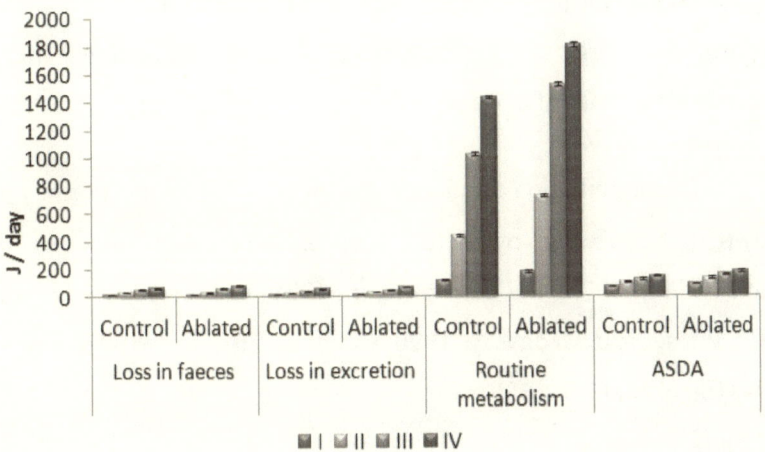

Figure 3.3b: Percent change in energy loss in faeces, excretion, routine metabolism and ASDA of ablated prawns over controls.

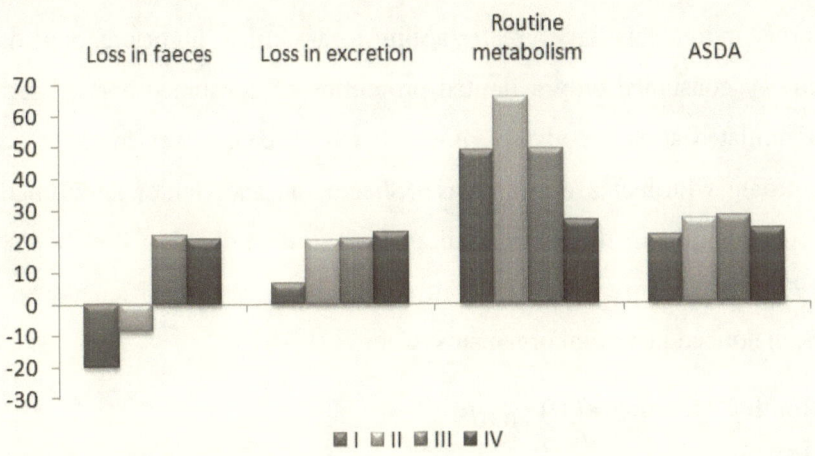

Both control and eyestalk ablated *M. malcolmsonii* showed greater proportion of consumed energy loss in excretion with increase in biomass (Figure 3.3a) is another evidence for energy utilization either by degradation of proteins in food and/or body proteins (Gerking, 1955; Cui and Wootton, 1988). Lower levels of percent energy loss in excretion after eyestalk ablation at each size group vis-a-vis control (Figure 3.3b) further indicate that prawns tend to lose smaller proportion of energy in excreation inorder to meet the additional energy demands arising out of eyestalk ablation.

The assimilation efficiency and gross conversion efficiency are known to be affected by factors such as temperature, biomass and ration (Elliott, 1976; Cui and Wootton, 1988; Keckeis and Schiermer, 1990; Wieser and Medgyesy, 1990; Carter and Brafield, 1991; Jobling, 1994; Hanel et al. 1996).

Decreased assimilation efficiency with increase in biomass in both the groups might explain increased energy loss in excretion with increase in biomass (Figure 3.2). It further indicates that although larger individuals have greater ability to assimilate higher amount of energy consumed only a limited proportion of consumed energy gets assimilated at each bodysize. Assimilation efficiency was found to be constant with increase in biomass (Sheerboom and Geldof, 1978) and food quantity in the pond snail, *Lymnaea stagnalis L.* (Sheerboom, 1978). But an inverse relation between feeding rates and efficiency has been noticed in several organisms (Calow, 1984).

Routine Metabolism (R_{rout}):

Bioenergetics modelling requires knowledge of the metabolic rates of an individual. Since direct measurement of energy loss is difficult, it is common to use oxygen consumption as an indirect measure of metabolism and use an oxycalorific equivalent to convert oxygen consumption into energy units (Brafield, 1985; Jobling, 1994). Knowledge of the oxygen requirements of aquatic animals is of great importance for aquaculture at all levels of physico-chemical conditions. Respiratory data provides a reliable basis for the calculation of carrying capacities, particularly of semi - intensive culture systems. Apparently maintenance metabolism (also measured as oxygen consumption) is the most important route for daily energy expenditure in most animals (Studier et al. 1975). The major factors affecting oxygen requirements of aquatic organisms are body weight, environmental temperature, salinity and diet (Jobling, 1994). Evaluation of various factors which might significantly affect oxygen consumption provides a measure of the importance of these intrinsic and extrinsic factors on the daily energy budget of species.

Results on changes in the energy values of routine metabolism in control and eyestalk ablated *M.malcolmsonii* are presented in figure 3.3a and the per cent change in energy allocated to routine metabolism between successive size groups is presented in Figure 3.3b. Routine metabolism increased significantly with increase in biomass in both control and eyestalk ablated prawns suggesting an increase in the costs of maintenance and routine physical activity thus elevating the total energy requirements of the animal. The effect of size on standard metabolism of an organism has been the focus of much research and conjecture (Zeuthen, 1953, 1970; Kleiber, 1961; Gordon, 1972). Several investigators reported increased oxygen consumption with increase in biomass in fish (Hughes, 1984; Oikawa and Itazawa, 1984; Itazawa and Oikawa, 1986). Diana (1983) found an increase in respiration with increase in age (0 to 3 years old) in the northern pike, *E. lucius* while Yamamoto (1991) observed increased oxygen consumption with increase in biomass in the carp, *Cyprinus carpio* (Linnaeus) under resting and normoxic condition. The relationship between metabolism and fish biomass follows a power law, values of the weight exponent varying between 0.7 and 0.9, with an extreme range of 0.5-1.0 (Brett, 1964; Fry, 1971). Several authors observed a linear relationship between the logarithm of metabolism and fish biomass (Brett, 1972; Brett and Grooves, 1979). The relationship between body weight and oxygen consumption in crustaceans such as *Penaeus schmitti* has been explained as early as in 1969 by Suarez and Xiques.

The per cent increase in energy allocated to routine metabolism between two successive size groups decreased with increase in biomass both in control and eyestalk ablated prawns. As there is a general decrease in the weight specific respiration rate with increase in size, the rate would be expected to decrease with development as reported by several authors (Mootz and Epifanio, 1974; Logan and Epifanio, 1978; Schatzlein and Costlow, 1978; Stephenson and Knight, 1980; Vernberg et al. 1981). However it may not hold true during development since the change in the life style may have a stronger effect on respiration rate than size due to changing oxygen demands (Zeuthen, 1953). Rate of oxygen consumption in several eyestalks less crustaceans has been reported to be higher than in intact individuals (Bliss, 1953; Raghavaiah et al. 1980; Koshio et al. 1992b).

Eyestalk ablation caused a significant increase in the respiratory rate (but not the rhythm) in post larvae and juveniles of *M. rosenbergii* which showed maximum oxygen consumption during the day (Rosas, 1991). Among other crustaceans elevation in whole animal oxygen consumption has been reported in the field crab *O. senex senex* (Reddy and Ramamurthi, 1987), *P. japonicus* (Chen and Chia, 1996) and *P. monodon* (Nan et al. 1995) upon eyestalk ablation. The level of routine metabolism in eyestalk ablated *M. malcolmsonii* of each size group was also found to be significantly higher (Student's t-test) compared to the respective controls indicating higher energy needs for metabolic activities related to maintenance or moulting along with tissue regeneration (Rosas et al. 1993). However decreased elevation in routine metabolism of eyestalk ablated prawns at later stages (size group IV) further provides evidence that prawns may not expend extra energy for maintenance during reproductive phase so as to facilitate gamate growth.

Percent of consumed energy allocated to routine metabolism also increased with increase in biomass in both control and eyestalk ablated prawns (Figure 3.3a) with the per cent energy allocated to routine metabolism being higher in eyestalk ablated prawns compared to the respective controls except in size group IV where both are similar (Figure 3.3b). This indicates that total maintenance energy demands of *M. malcolmsonii* increase with increase in biomass and eyestalk ablated prawns have to allocate higher proportion of consumed energy for maintenance and / or routine metabolism.

Apparent specific dynamic action (ASDA):

The length of time for which consumption of food exerts its influence upon heat production depends upon many factors: Chief amongst them are the quantity and quality of food (Brett and Grooves, 1979; Beamish and Trippel, 1990; Carter and Brafield, 1992) and water temperature (Cho et al. 1982). Kleiber (1961) used the term "heat increment" to these energy changes instead of Specific Dynamic Action (SDA) formerly applied to denote changes accompanying protein deamination. A major part of SDA could be thought to represent the metabolic cost of growth (Jobling 1985; Wieser, 1994) SDA is only one component of "apparent SDA", a term coined by Beamish (1974) to describe the total postprandial rise in metabolic rate.

Results on changes in energy values of apparent specific dynamic action (ASDA) in control and eyestalk ablated *M. malcolmsonii* are presented in figure 3.3a. Significant increase in ASDA with increase in biomass has been noticed in both control and eyestalk ablated prawns. The energy loss associated with feeding can be attributed to several causes including excited locomotor and other incidental activities, mastication, movement of food through gut, enhancement of digestive secretions, digestion and absorption of food and biochemical transformation of absorbed materials (Jobling, 1994). The latter is especially associated with the metabolism of proteins and amino acids (Borsook, 1936; Nelson and Cox, 2000) but also includes release of energy accompanying lipid and carbohydrate metabolism. Energy expenditure for ASDA is found to be influenced by meal size (Muir and Niimi, 1972), environmental temperatures (Brett, 1976), biomass (Beamish, 1974) and diet (Cho et al. 1976; Smith et al. 1978; Tandler and Beamish, 1980).

M. malcolmsonii was provided with dry pelleted diet consisting 38-42% protein, 5-8% fat and 5% fibre. Hence ASDA obtained can be directly attributed to the utilization of proteins. Further the fact that a significant increase in ASDA with increase in biomass in both control and experimental groups (Figure 3.3a) was parallel with ingestion levels suggests higher rates of digestion and mobilization of materials through gut. Apparent heat increment depends on the quantity, quality and balance of the dietary energy components (Harpar, 1971; Buttery and Annison, 1973) and on the nutritional status of the animal (Smith, 1981; Lied et al. 1982). Meal size directly affected apparent heat increment for a wide variety of fish species fed with a broad spectrum of natural and formulated diets (Beamish, 1974; Vahl and Davenport, 1979). The energy requirements of growth in largemouth bass were found to be positively related to biomass when meal size is expressed relative to fish biomass (Niimi and Beamish, 1974). Comparable studies in fish (Beamish, 1974; Tandler and Beamish, 1979) and prawns (Sindhu Kumari and Pandian, 1987; Du Preeze et al. 1992) provided further informtion required to predict a positive relationship between SDA and biomass / amount of food consumed. Carefoot (1990) reported that SDA asymptotically increased with ration and had significant positive effect on biomass in the supralittoral isopod, *L. palassi*.

Rate of increase in energy allocation to ASDA decreased gradually between successive size groups in both control and eyestalk ablated prawns with increase in biomass. Decreases in the rate of increase in ASDA between successive size groups further strengthen the view that the maximum O_2 uptake relative to basal metabolism declined curvi-linearly with increase in biomass (Tandler and Beamish, 1981). This implies that the absolute amount of energy utilized for ASDA depends on the amount of food ingested which inturn is influenced by body size of the animal. Allocation of higher levels of energy to ASDA by eyestalk ablated prawns compared to the respective controls (Figure 3.3b) could be explained on the basis of higher levels of food ingestion (Figure 3.1a). Decrease in absorption efficiency and assimilation efficiency in *M. malcolmsonnii* have been shown to suggest that the capacity of utilizing ingested food may be a function of body size and / or age as explained by Smith (1973). This is reflected further by a decrease in percent elevation in ASDA in eyestalk ablated *M. malcolmsonii* of size group IV which exhibited maximum food intake compared to their respective controls (Figure 3.1a). Specific dynamic effect was found to be similar between normal and eyestalk ablated crayfish (Sierra et al. 1997).

In the present study although no significant difference was found in the proportion of ingested energy allocated to ASDA between control and eyestalk ablated prawns of size group I, II, and III, significantly lower proportion of energy has been allocated to ASDA in eyestalk ablated prawns of size group IV. The most frequently discussed aspect of energy partitioning in poikilothermic animals has been the trade-off mechanism (Townsend and Calow, 1981). Production is mainly channeled to increase animal biomass (and therefore energy content) or to the formation of gametes and embryos. Thus a major portion of production is usually associated with growth or with reproduction (Brafield and Llewellyn, 1982). Induction of maturation or breeding through eyestalk ablation is well known among adult crustaceans (Tan - Fermin, 1991; Chaves, 2000). Since gamete maturation is an energy demanding process, conservation of energy from other life-history traits is necessary to attain sexual maturity. Hence reduction in the proportion of injested energy allocated to ASDA in prawns of size group IV (Figure 3.3a) which have attained the biomass of (15-20g) sexual maturity, provide indications of the 'trade-off mechanism' in *M. malcolmsonii* which is required to maximize growth - the phenomenon which has been well studied in higher animals such as fishes (Handy et al. 1999).

Scope for growth (SFG):

The term "Growth" is broadly used to imply a change in quantity such as mass, protein or other chemical constituents. Growth may relate to the reproductive ability of cells as a function of cell differentiation or to the expenditure and distribution of energy as a function of food consumed under various environmental conditions (Beamish and Trippel, 1990). The energy available for growth and gamet production is called "scope for growth" (SFG) was calculated by Warren and Davis in 1967 for each periwinkle as the difference between the rates of energy gain (food absorbed) and energy loss (respiration and excretion). Some models on the evolution of life cycles assume that the physiological basis of tissue production in reproduction is the same as that in somatic growth (Sebens, 1979; Calow, 1981 a,b). Alternatively, others draw attention to the fact that reproduction may begin after growth has ceased and it therefore probably involves different metabolic processes (Calow et al. 1979; Calow, 1981a). However, in the present study, energy available for both somatic and reproductive growth has been assessed together.

Growth of an individual within a population is often highly variable. The causes vary involving genetic, social and environmental factors. Animal growth involves cells, tissues, organs and the organism as a whole and each level has specific inhibitory and stimulatory factors acting on the organs (Needham, 1964). Variation in growth of an aquatic animal is also of practical interest, as it influences the harvest of fisheries and aquaculture system (Raanan et al. 1991). Animals ingest macromolecules and break them down by digestion into monomers, which are assimilated from the gut. Some of them are used in respiration while some others are assembled into polymers which comprise animal production (Brafield and Llewellyn, 1982). Calow (1977) has calculated the energy required to build polymers from monomers to produce animal tissue. Both growth and reproduction require considerable energy and materials, and consequently they are antagonistic (Bohlken and Josse, 1982; Joosse and Geraerts, 1983).

Increased scope for growth with increase in biomass in both control and eyestalk ablated M. malcolmsonii (Figure 3.1a) suggest that the amount of energy available for growth (somatic / reproductive) and /or exuvia production increases with increase in biomass. The energy requirements for growth in largemouth bass were found to be positively correlated to body weight when meal size is expressed relative to fish biomass (Niimi and Beamish, 1974).

However the rate of increase between successive size groups decreased gradually in both control and eyestalk ablated prawns with increase in biomass. Von Bertalanffy (1938) made one of the first attempts to explain animal growth rates. He considered that the growth rate was a function of two processes, synthesis and degradation and that these two are dependent on body size and differences in growth rate were either attributed directly to differences in dissolved oxygen regime in the external medium or indirectly to differences in feeding rates. Since dissolved oxygen is available to saturation levels (Table 4) in the present study, any variation in growth rates could only be attributed to feeding levels and / or energy expenditure.

Figure 3.4: Energy budget for control and ablated *M. malcolmsonii*.

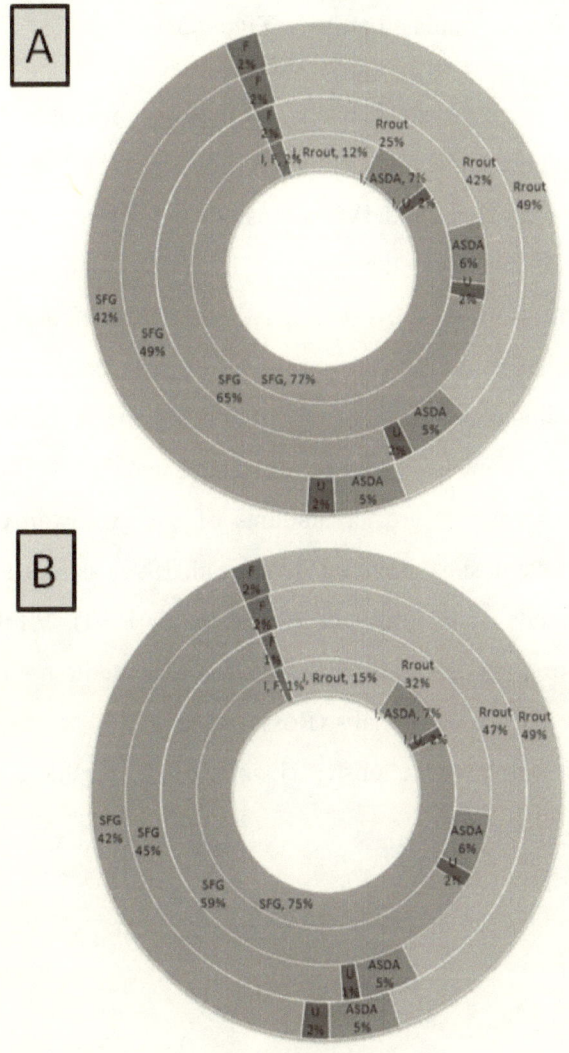

Levels of scope for growth are higher in eyestalk ablated prawns at each size group compared to the respective controls (Figure 3.3b). Greater elevation at later stages (size III and IV) further emphasise the influence of eyestalk ablation on growth potential. Although percentage of energy potentially available for growth decreased with increase in biomass in both control and eyestalk ablated prawns, no clear trend has been visualised in the change between control and eyestalk ablated prawns (Figure 3.4).

The proportions of energy potentially available for growth (Somatic, reproductive and exuvia production) in eyestalk ablated prawns (Group - I, II and IV) compared to their respective controls indicate a change in the pattern of energy allocation in M. malcolmsonii as result of eyestalk ablation thus affecting the amount of energy potentially available for growth. For instance eyestalk ablation initiates vitellogenesis (Chaves, 2000) and the energy needed for it is derived from the ingested energy. Destalked female crayfish P. clarkii was found to channel greater amounts of energy towards scope for growth than the destalked male (Sierra et al. 1997) as a result of lesser energy demands of maintenance (Rosas et al. 1993). While eyestalk ablation caused an increase in scope for growth in both male and female pink shrimp, P. notialis (Rossas et al. 1993) the same resulted in fast growth in juvenile lobster H. americanus (Koshio et al. 1992a).

Chapter 4: Energy metabolism

Carbohydrate Metabolism

Carbohydrate metabolism is essentially the metabolism of glucose and substances related to glucose. Glucose occupies the central position of carbohydrate metabolism in an organism, representing complex groups, sequences and cycle of reactions which integrate at various points the reactions concerned with metabolism of lipids and proteins as these molecules serve the source of carbon in the synthesis of cellular components (Nelson and Cox, 2000). Glycogen is the chief carbohydrate present in tissues, while glucose is of the blood and other body fluids. Glycogen, a storage carbohydrate, is reversibly converted to blood glucose and normally serve to maintain blood sugar level when supply of carbohydrate from intestinal absorption is inadequate. Glycogen breakdown into glucose is governed by the extrinsic and intrinsic factors which also controls the physiology of an organism.

Carbohydrate metabolism essentially constitutes two segments: synthesis of carbohydrates which includes –glycogenesis and gluconeogenisis and catabolism which includes-glycolysis, glycogenolysis, protease pathway and Krebs cycle. The Catabolic pathways not only fulfil the needs of energy demands but also supply the amphibolic intermediates and reduced nucleotides (NADPH) required for proteins and lipid metabolism (Nelson and Cox, 2000). The mechanism by which glycogen is synthesized and broken down in tissues is initiated by glycogen phosphyorylase enzyme. The process of glycolysis in tissues commences with the onset of glycogen breakdown and the glucose released is fragmented in to three carbon compounds, pyruvic acid and lactic acid by a series of enzymes under anaerobic conditions. The end products lactate and pyruvate are interconvertible by the enzyme lactate dehydrogenase (LDH). Pyruvate undergoes oxidative decarboxylation by pyruvate dehydrogenase to provide acetyl-CoA.

Acetyl-CoA is an essential substrate for Krebs cycle, which generates reduced nucleotides for the ultimate generation of ATP molecules through electron transport system. Amphibolic intermediates formed in the Krebs cycle may be channelled into amino acid or fatty acid synthesis. Channelling of carbohydrate precursors into energy yielding reactions or synthetic reactions depends on the biochemical makeup of an organ system concerned or physiological alterations attendant on changed environmental conditions (extrinsic or intrinsic). Pyruvate utilization for energy requirements is tissue specific nd varies according to environmental conditions imposed on the animal. Since biological system has some amount of flexibility, animals use this capacity to divert metabolic pathways to an alternate source so as to synthesize energy to overcome the energy crisis created by the stress.

The present investigation was carried out to study various segments of carbohydrate metabolism in different tissues of control and eyestalk ablated M. Malcolmsonii. Figure 4.1a and 4.1b depict the results on total carbohydrate (TCHO) and glycogen content of the hepatopancreases and muscle and hemolymph glucose in the four size groups of controls and eyestalk ablated *M. Malcolmsonii*.

The results of the present investigation clearly demonstrate an overall shift in the carbohtdrate metabolism characterised by the time dependent alterations in metabolites and associated enzymatic activities of the hepatopancreas, muscle and hemolymph of *M. Malcolmsonii*. TCHO content was higher in hepatopancreas than in the muscle of control prawns, which could be attributed to the fact that hepatopancreas is an active site for synthesis and storage of glycogen as reported by Huggins (1966). Since hepatopancreas is the metabolic centre (Chang and O'Connor, 1983), the presences of non-glycogenic carbohydrates also contribute to TCHO level. Most of the crustaceans are known to have oligosaccharides like maltose, trehaloes and amino sugars in hepatopancreas (Chang and O'Connor, 1983). Muscle comes next to the hepatopancreas with regard to concentration of TCHO and glycogen indicating that the muscle tissue also can store these macro molecules. Schwoch (1972) observed that the muscle, hepatopancreas and integumentary tissue of the crayfish, Orconectes limosus, act as synthesis and storage organ for maltose and trehalose. Thus accumulations of non-glycogenic sugars contribute for TCHO content similar to hepatopancreas.

The resuts also demonstrate that total carbohydrate and glycogen contents in both hepatopancreas and muscle increased with increase in biomass of prawns indicating that the ability to synthesize and store these macromolecules increase with increase in biomass. Increase in glycogen content with increase in biomass has also been reported earlier in other freshwater invertebrates such as amphipods (Quigley et al. 1989), rotifers (Guisande and Serrano, 1989), cladocerans (Goulden and place, 1990) and stonefly larvae (Meyer, 1990).

Upon eyestalk ablation both TCHO and glycogen contents decreased significantly both in hepatopancreas and muscle of all the size groups of M. malcolmsonii compared to the respective controls (Figure 4.1b) indicating possible utilization of TCHO and glycogen to overcome the stress effect. Similar results have also been reported in *S. serrata* (Ramana Rao et al. 1991) and in other crustaceans such as *M.monoceros* (Surendranath et al. 1992) after unilateral eyestalk lation.

Figure 4.1a: Total carbohydrate in control and eyestalk ablated *M. malcolmsonii* of four different size groups.

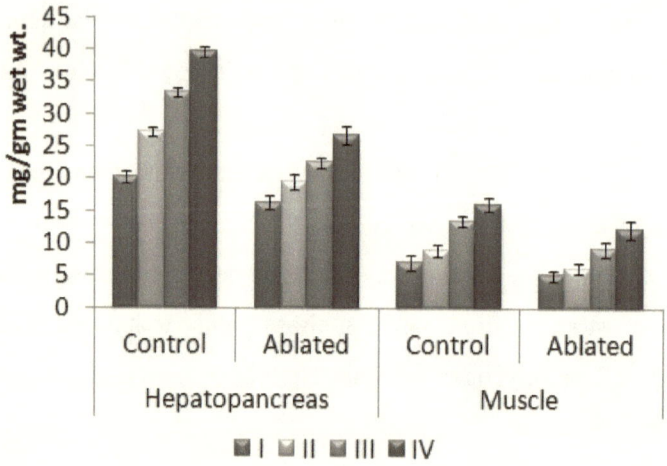

Figure 4.1b: Percent change in total carbohydrate of ablated prawns over controls in four different size groups.

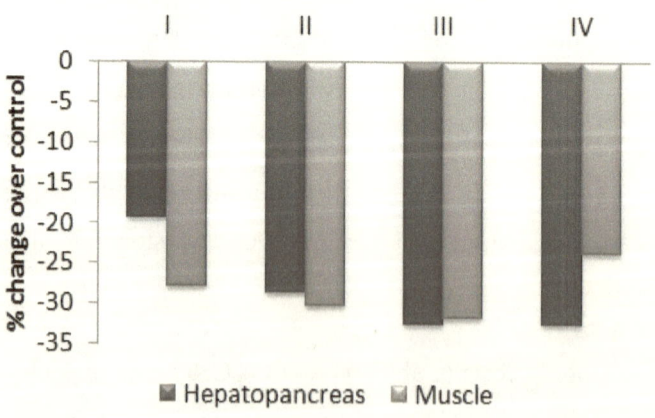

It also suggests that the absence of the eyestalk hormones seem to impose some additional energy demands related to moulting and osmoregulation. Bliss in 1968 noticed a rise in hemolymph osmotic pressure, followed by a rapid uptake of water during the late premoult period which is used to aid in shedding the old exoskeleton and stretching the new exoskeleton after ecdysis. Shoot-up in respiratory rate has also been reported in eyestalk ablated prawns (Rangneker and Madhyastha, 1969; Koshio et al. I992b). Consequently there should be higher breakdown of glycogen and an increase in the hemolymph glucose. Much of the tissue glycogen and hemolymph glucose appeared to be chanelled into energy expending processes associated with moulting and cuticle formation (Hornung and Stevenson, 1971). A similar trend was also seen in eyestalk ablated M. malcolmsonii, particularly regarding tissue glycogen and hemolymph glucose levels (Figure 4.2a). This should naturally impose drain on animal carbodrate reserves. It was also found that utilization of glycogen reserves seems to be more in hepatopancreas than in muscle subscribing to the view that hepatopancreas is the main metabolic centre.

Hemolymph glucose levels also increased signiflcantly with an increase in biomass of M. malcolmsonii signlfying the importance of greater energy demands of larger individuals (Figure 4.2a). The significant increase in hemolymph glucose level in eyestalk ablated prawns (Figure 4.2b) along with a decrease in tissue glycogen (Figure 4.2b) clearly suggest breakdown of glycogen and its conversion to glucose. Elevated levels of hemolymph glucose have been reported in destalked mature crab *Scylla serrata* (Deshmukh, 1968) and immature *Scylla serrata* (Ramana Rao et al. I991), *Paratelphusa jcaguemontil* (Rangneker et al. 1971) *Ocypod platytarsis* and *Varuna litterata*

(Parvathy, 1972; Madhyastha and Rangneker, 1976). Similarly Surendranath et al. (1992) reported an increase in hemolymph glucose and decrease in muscle glycogen in eyestalk ablated prawn M. monoceros.

Figure 4.2a: Glycogen and hemolymph glucose in control and eyestalk ablated *M. malcolmsonii*.

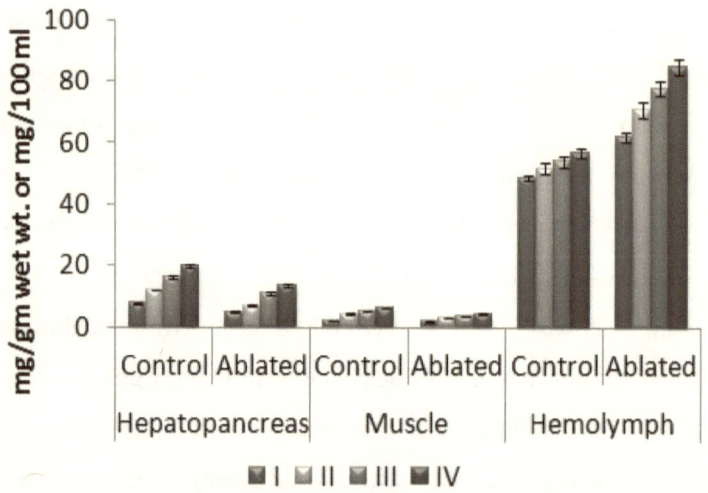

Figure 4.2b: Percent change in glycogen and hemolymph glucose of ablated prawns over controls in four different size groups.

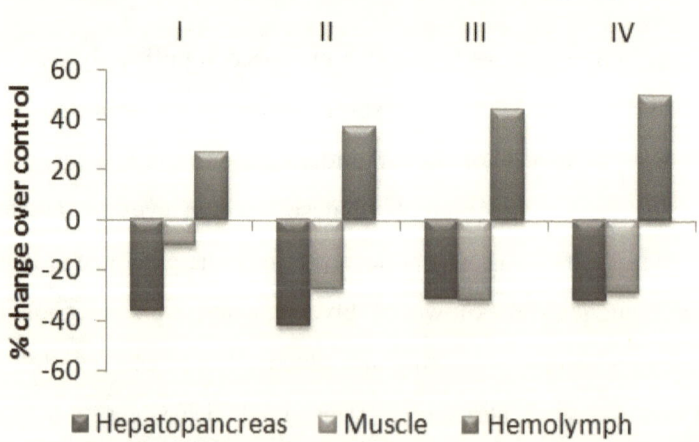

Glucose is the major circulating carbohydrate in crustaceans (Telford, 1968; Dall, 1975). It serves as a precursor for the synthesis of oligosaccharides and glycogen in crustaceans (Meenakshi and Scheer, 1961). It has been reported by several workers that the hemolymph glucose levels are influenced by various conditions such as moult cycle (Telford, 1968) and reproductive cycle (Dean and Vernberg, 1965). A rise in hemolymph glucose titre is a classical response exhibited by crustaceans to physiological disturbances (Dall, 1975).

Hyperglycaemia has been documented as a second stress response in most fish species (Wedemeyer, 1972; Carmichael et al. 1984; Reobertson et al. 1988). Several crustaceans also showed increased hemolymph glucose levels when subjected to various stressful conditions. For instance significant increase in hemolymph glucose levels have been observed in the red swamp crayfish Procambarus clarkia (Reddy et al. 1994) and the fiddler crab Uca pugilator (Reddy et al. 1996) exposed to hypoxia or cadmium and in Orconectes limosus (Santos and Keller, 1993) subjected to hypoxia, starvation and elevated temperatures. Homeostasis of hemolymph glucose was found to occur by an interaction between crustacean hyperglycemic hormone (CHH) release and carbohydrate metabolites in the hemolymph. During increased glycolytic flux, lactate may archestrate the release of CHH by a positive feedback mechanism. The hormone then stimulates glycogenolysis thus increasing glucose availability.

Lactate Dehydrogenase (LDH):

Tissue lactate and pyruvate contents constitute the interconvertible terminal products of glycolysis. Their interconversion is mediated by the enzyme lactate dehydrogenase (LDH) which requires NAD* / NADH$^+$ as coenzymes. The NAD* dependent LDH catalyzes the conversion of lactate to pyruvate. Results on the activity levels of LDH in control and eyestalk ablated M. malcolmsonii of four different size groups were presented in figure 4.3a. The results show that LDH activity in both muscle and hepatopancreas of *M.malcolmsonii* increased with increase in biomass (Figure 4.3a) indicating increased rate ofconversion of lactate to pyruvate or vice versa. In general variations in the activity level of tissue specific LDH indicate differences in the physiological and metabolic status of the respective tissues. The higher level of LDH activity of the hepatopcreas noticed in the present study may be due to greater metabolic functions that this organ can perform. Since hepatopancreas of crustaceans is more analogous to the liver of vertebrates, where lactate is converted to pyruvate for metabolic purposes, presence of higher level of LDH activity in hepatopancreas over muscle is an expected phenomenon.

Upon eyestalk ablation there was a significant increase in LDH activity in the muscle and hepatopancreas of *M. malcolmsonii* of all the four size groups (Figure 4.3b). This observation is in agreement with the reports of Reddy (1990) who has noticed increased LDH activity in the tissues of eyestalk ablated crab O. senex senex. Reports indicate eyestalk ablation to cause changes in osmoregulation (Kamemoto, 1976; Nagabhushanam and Jyothi, 1977) and acid-base balance (Truchot, 1983). Therefore the pyruvate content may indirectly facilitate the maintenance of osmoregulatrion, while lactate content might function in maintaining the acid - base balance (Mantel and Farmer, 1983; Rosas et al. 2001).

Figure 4.3a: LDH activity levels in control and eyestalk ablated *M. malcolmsonii*.

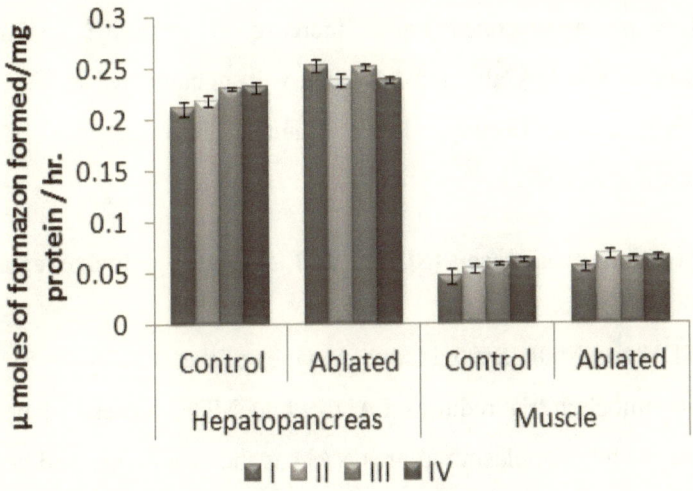

Figure 4.3b: Percent change in LDH activity levels of ablated prawns over controls.

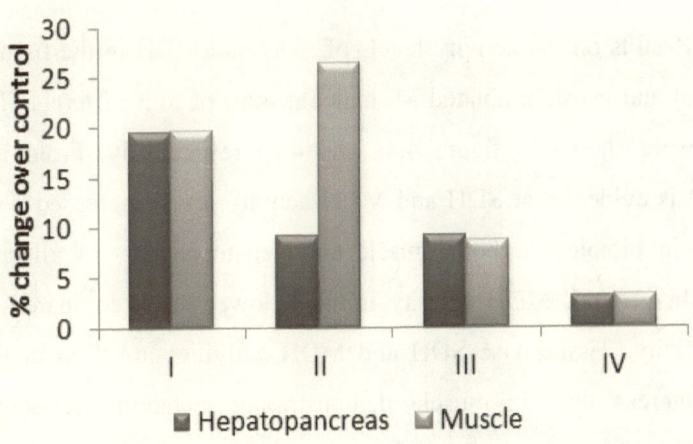

Eyestalk ablation results either in moulting or in reproduction (Sagi et al. 1997; Rotllant et al. 2001) and thus significantly increases the energy demands of crustaceans (Adiyodi and Adiyodi, 1970). Hence in the present study increased activity levels of LDH upon

115

eyestalk ablation could be meant to keep the pyruvate concentrations at a higher level so as to ensure energy production through oxidative metabolism on a sustained basis. Increase in oxidative enzyme activities upon eyestalk ablation has been well documented earlier in several crustaceans (Reddy, 1990; Ramana Rao et al. 1991; Surendranath et al. 1992).

Succinate dehydrogenase (SDH) and Malate dehydrogenase (MDH):

SDH and MDH are oxidative enzymes of the TCA cycle. SDH present in mitochondria requires FAD, while MDH present in both mitochondria and cyboplasm requires NAD as the coenzyrnes and both these enzymes contribute to the synthesis of ATP. Therefore these enzyme systems have also been taken up for investigation in the present study.

Results on the activity levels of SDH and MDH in the tissues of control and eyestalk ablated M. malcolmsonii of four different size groups were shown in figure 4.4a and 4.5a respectively. From the results it is evident that SDH and MDH activity levels increased with increase in biomass in both muscle and hepatopancreas of all size groups. In general, MDH activity is much lower when compared to SDH activity. Tissue-wise SDH and MDH activities are high in the hepatopancreas than in muscle demonstrating metabolically active nature of hepatopancreas over muscle while low activity levels in muscle denote its glycolybic nature (Hu, 1958; Huggins and Munday, 1968) many reports confirmed higher oxidative metabolic activity of crustacean hepatopancreas (Huggins and Munday, 1968; Chang and O' Connor, 1983).

Figure 4.4a: SDH activity levels in control and eyestalk ablated *M. malcolmsonii*.

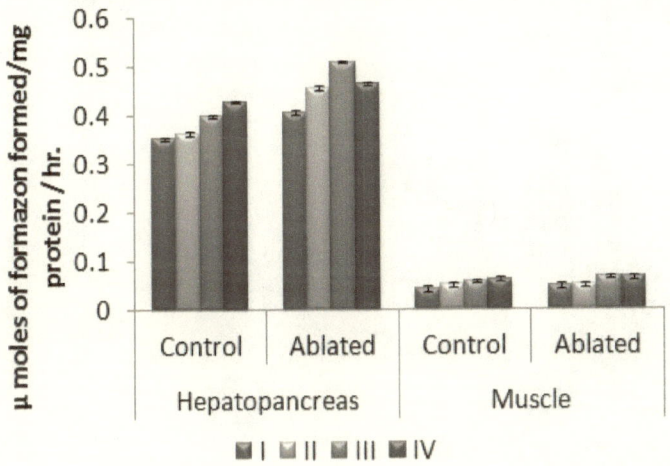

Figure 4.4b: Percent change in SDH activity levels of ablated prawns over controls.

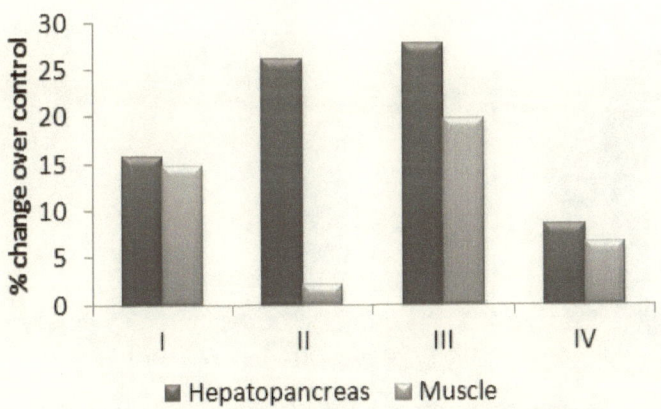

Upon eyestalk ablation, there was a significant increase in SDH and MDH activity levels in muscle and hepatopancreas of *M. malcolmsonii* of all the four size groups (Figure 4.4b. and 4.5b). Elevated levels of oxidative enzyme activities observed in the present study are in conformity with the observations of Zerbe et al. (1970) in *Astacus leptodactylus* and Reddy (1990) in crab *O. senex senex*.

Figure 4.5a: MDH activity levels in control and eyestalk ablated *M. malcolmsonii*.

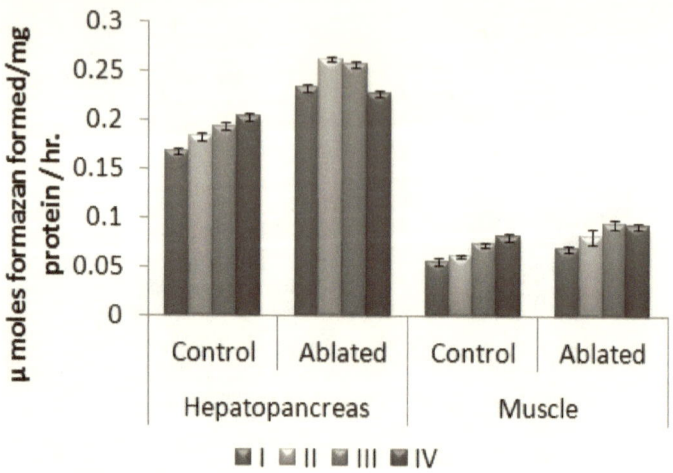

Figure 4.5b: Percent change in MDH activity levels of ablated prawns over controls.

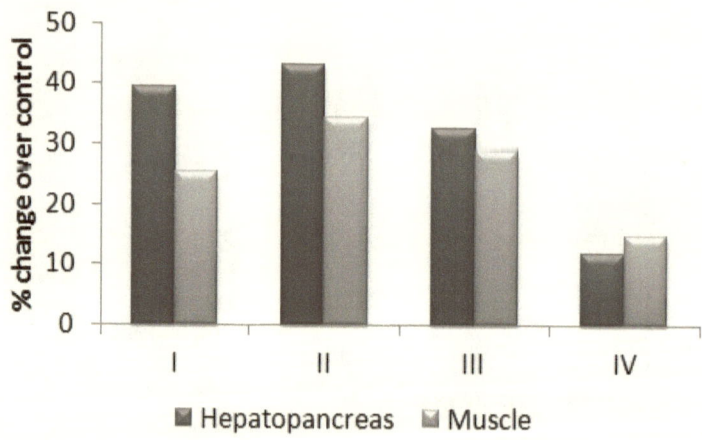

The elevation of SDH and MDH in hepatopancreas and muscle, clearly demonstrate elevated rate of Krebs cycle activity. This could be an indication of enhancement in oxidative phosphorylation towards ATP synthesis and is a biochemical background presumably favourable for the generation of more ATP, the "celluxar energy currency". It also indicates that absence of eyestalks imposes a stress on these tissues favouring more production of ATP so as to meet extra energy demands arising out of altered metabolic and physiological conditions.

Protein metabolism:

Proteins, by far, are the most important group of macromolecular chemical substances which occupy a pivotal place in both structural and dynamic aspects of living systems (Murray et al. 2000). Further the catabolic products of proteins appear in the form of direct nitrogenous substances which are known to play a key role in several physiological processes of animals (Nelson and Cox, 2000). Metabolic response in proteins is considered to be one of the principal physiological events involved in the compensatory mechanisrn in terms of homeostasis under any stress condition (Assem and Hanke, 1983). Protein synthesis and degradation are reflected by changes in protein composition (Robert and Bocyuen, 1974). Proteins of animal tissues are recognized to exist in a dynamic steady state undergoing continuous synthesis and degradation (Goldberg, 1974).

Tissue proteins undergo a continuous process of renewal, referred to as "turnover". Protein concentrations are determined by the rates of degradation and synthesis both being regulatory in nature (Segel et al. 1976). Proteins must be continuously supplied to the organisms for growth and are to be maintained at constant levels (Dunlop et al. 1978; Hershko and Ciechanover, 1982). Further changes in protein concentrations probably reflect numerous physiological changes going on in the organism during growth period (Anne Bond et al. 1993). Unlike in mammals protein is both a building block and an energy source in crustaceans (Pandian, 1989). The study of variations in energy storage in the form of protein would thus help understand the ecology and overall economy of the species (Sreedhar and Radhakrishnan, 1995).

Total Protein (TP):

The results on total protein content in control and eyestalk ablated prawns of four different size groups were depicted in figure 4.6a and 4.6b. The total protein content in control prawns of four different size groups was high in muscle followed by hepatopancreas and hemolymph. Variations in total protein content in the tissues of control prawn reflect the diversity of structure and functions of these tissues. High total protein content in the muscle (~73% of dry weight) has been reported earlier in the brown shrimp Penaeus aztecus (Shewbart et al. 1972). It is only natural that the muscle, which provides the structural framework, contains higher amounts of total protein than hepatopancreas which serves as the main metabolic centre.

Figure 4.6a: Total protein in control and eyestalk ablated *M. malcolmsonii*.

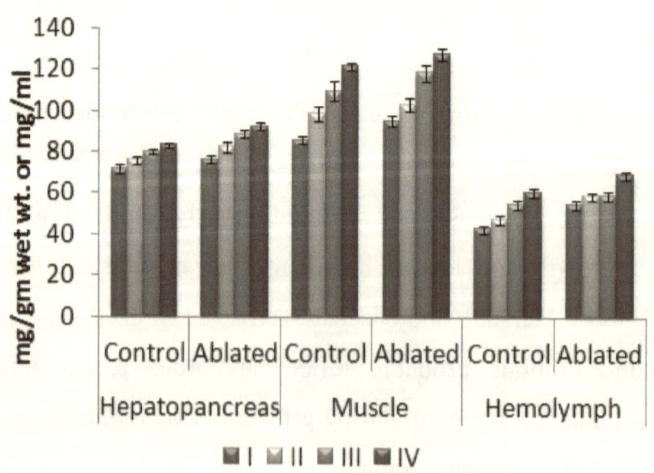

Figure 4.6b: Percent change in total protein of ablated prawns over controls.

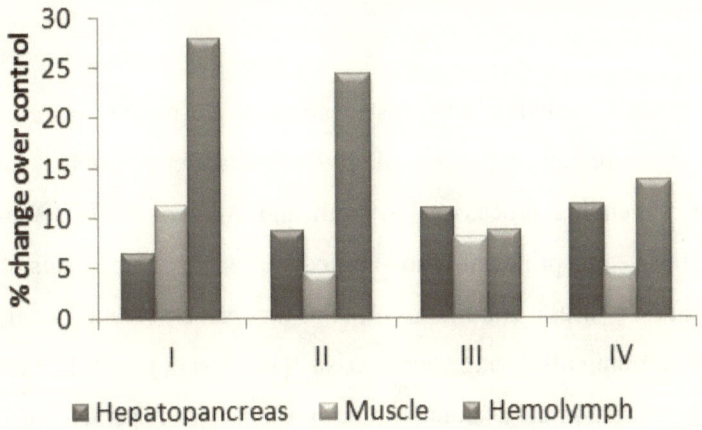

The results also depict that total protein content increased with increase in the body size of prawns with the highest protein content being recorded in IV size group of prawns (Figure 4.6a). This is natural because growth is associated with tissue accumulation, the main component of which is protein. This could especially be so in prawns because total protein in crustaceans serves both as a building block and as an enerry source (Ennis, 1972; Pandian, 1989; Gendron et al. 2001). Increase in energy storage reserves in the form of protein with increase in biomass has been reported in several aquatic invertebrates (Quigley et al. 1989; Meyer, 1990; Larson, 1991, Reddy et al. 1992).

Upon eyestalk ablation there was a significant increase in total protein content of hepatopancreas, muscle and hemolymph in all the four size groups of *M. malcolmsonii* (Figure 4.6b). The above observation indicates a progressive change in tissue as a preparatory attempt to adopt to increased enerry demands. Some studies reported the enhanced protein degradation (Adiyodi, 1968; Raghavaiah et al. 1980; Reddy, 1990). On the other hand McWhinne (1962) reported a significant increase in hemolymph protein levels in the eyestalk ablated crayfish Orconectes uirilis (Hagen) as has been reported in the present study.

Eyestalk ablation in crustacean species results in either moulting or reproduction (Sagi et al. 1997; Rotllant et al. 2000) which is energy expending processes (Adiyodi and Adiyodi, 1970). While carbohydrates are principal and immediate energy precursors for animals during stress condition, proteins act as emergency energy precursors for animals during energy crisis (Gendron et al. 2001). In the present investigation the accumulation of proteins was observed both in hepatopancreas and muscle in all four size groups. While Ramamurthi et al. (1981) recorded incorporation of labelled leucine and aspartic acid into hepatopancreas and chelate leg muscle of bilateral eyestalk alated crab O. senex senex, Jayasundaramma and Ramamurthi (1988) noticed the incorporation of labelled leucine alone in hepatopancreas and gonad of the same species. Recently Chaves (2001) noticed incorporation of (^{35}S) methionine into ovary and hemolymph proteins of eyestalk ablated M. rosenbergii and Palaemonetes kadiakensis (Rathbun) indicating increased protein synthesis. A significant increase in protein level of hemolymph, which transports enerry rich substance from one part of the body to the other, could be attributed to the translocation and transportation of protein either from hepatopancreas or from muscle into the hemolymph.

Free Amino Acids (FAA):

Figure 4.7a and 4.7b presents data on free amino acid (FAA) content in the hepatopancreas, muscle and hemolymph of control and eyestalk ablated M. malcolmsonii of various size groups. It was clear that FAA content of all the three tissues increased with increase in biomass and that the FAA content in the muscle of control M.malcolmsonii is higher followed by hepatopancreas and hemolymph reflecting the positive relationship between muscle total protein (Figure 4.6a) and FAA content (Figure 4.7a). Similar results have also been reported in a few crustaceans (Camien et al. 1951; Schoffeniels and Gilles, 1970; Claybrook, 1983). Gerard and Gilles (1972) reported that muscle of the blue crab, Callinecteus sapidus contains high FAA content compared to hepatopancreas. Similarly Claybrook (1983) found that the free amino acid content is relatively high in the muscle, intermediate in hepatopancreas and low in hemolymph and gill.

Figure 4.7a: Free amino acid content in control and eyestalk ablated *M. malcolmsonii*.

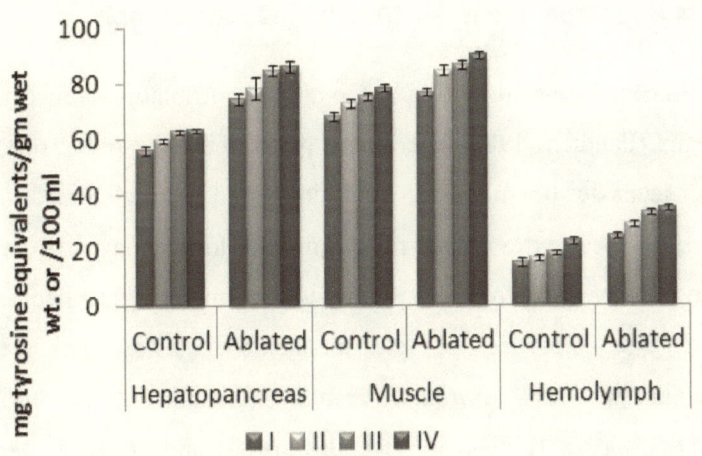

Upon eyestalk ablation there was a significant increase in FAA content of the muscle, hepatopancreas and hemolymph of *M. malcolmsonii* of all the four size groups (Figure 4.7b). A significant increase in FAA content could be due to the degradation of muscle and hepatopancreatic protein. Claybrook (1983) reported that total free amino acid pool is very high in crustaceans and, it is quite likely that a part of this pool may be utilized to meet the energy demands of moulting.

Figure 4.7b: Percent change in free amino acid content of ablated prawns over controls.

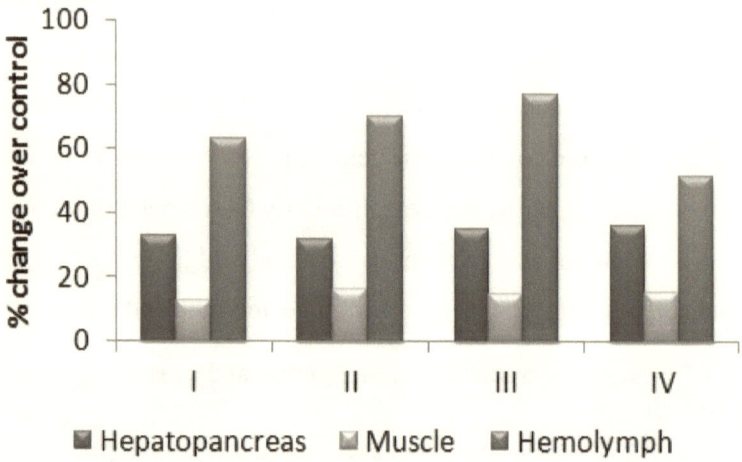

Since proteins are said to be the major contributors of energy in crustaceans (Pandian, 1989), the amino acids being the precursors of proteins, seem to be mobilized and converted into gluconeogenic precursors to meet extra enerry requirements. However a part of this pool may also be used for energy biosynthesis through gluconeogenic pathway. The presence of gluconeogenic pathway has been documented in *Crabs Carcinus meanus* (Chaplin et al. 1967), *Orconectes Iimosus* (Urich, 1967; Gilles, 1969) and *O. senex senex* (Raghavaiah et al. 1980; Reddy, 1990). Increase in FAA content upon eyestalk ablation also reported in the Crab *S. serrata* (Radhika et al. 1988). On the whole increase in FAA levels (Figure 4.7a) indicate potential elevation of proteolysis in *M. malcolmsonii* after eyestalk ablation.

Protease:

Protease catalyzes the breakdown of proteins resulting in formation of total free amino acids. The results clearly demonstrate protease activity increased with increase in biomass (Figure 4.8a). In the control prawns protease activity was relatively high in the hepatopancreas followed by muscle (Figure 4.8a). Hepatopancreas being the metabolic centre in prawns (Adiyodi and Adiyodi, 1970), the high metabolic potential of this tissue should favour both synthesis and breakdown of the products. Muscle mostly participates in physiological functions and that its metabolic needs are rather limited confined to structural and functional dynamics. Upon eyestalk ablation there was a significant increase in protease activity in both hepatopancrease and muscle of *M.malcolmsonii* of all the four size groups (Figure 4.8b).

Figure 4.8a: Protease activity levels in control and eyestalk ablated *M. malcolmsonii*.

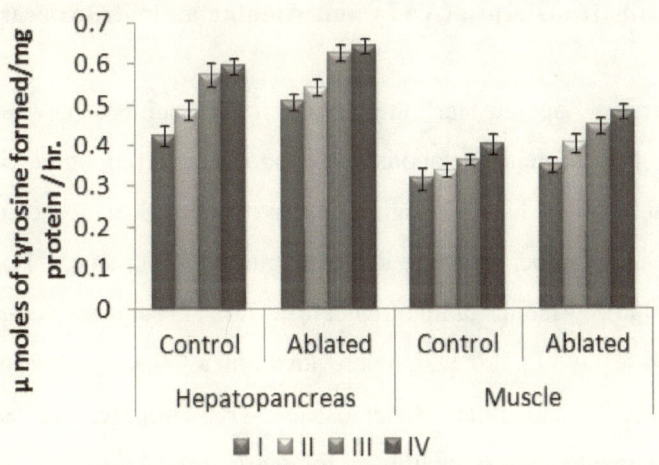

Pandian (1989) reported that crustaceans utilize proteins as an energy source rather than carbohydrates and fats. Zandee (1966) is of the opinion that variations in FAA in crustacean tissues may be considered as an indication of their use in the process of energy production. Therefore, an increase in tissue protease activity along with enhanced FAA appears to be in consonance with the metabolic needs of the eyestalk ablated *M.malcolmsonii*.

Figure 4.8b: Percent change in protease activity levels of ablated prawns over controls.

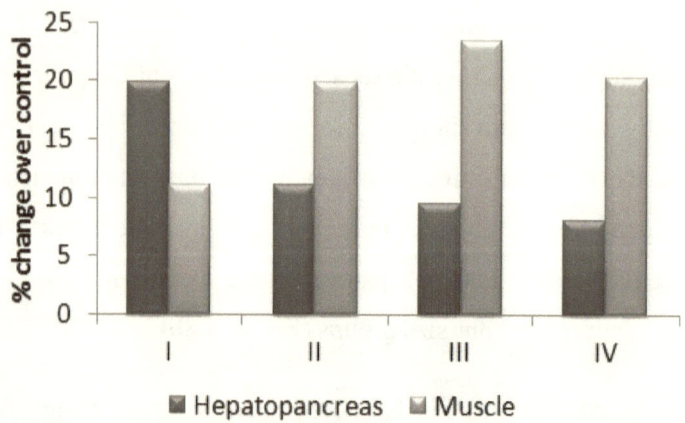

Aspartate aminotransferase (AAT) and Alanine aminotransferase (AlAT):

Aminotransferases operate at the 'cross over points' between carbohydrate and protein metabolism by inter converting stlategic 'cross over metabolites' like o-ketoglutarate, pyruvate and oxaloacetate on one hand and alanine, aspartate and fumarate on the other (Nelson and Cox, 2000). Alanine aminotransferase (AlAT) and aspartate aminotransferase (AAT) are widely prevalent in all the crustaceans (Hartenstein, 1970) and other animal tissues. Transaminases convert amino acids into keto acids to be utilized for energy production.

The results on the activity levels of AAT and AlAT in control and eyestalk ablated M. malcolmsonii of four different size groups were presented in Figure 4.9a and 4.10a respectively. From the results it is clear that AAT and AlAT activities of both tissues increased with increase in biomass (Figure 4.9aand 4.10b). This can be attributed to the role of transaminases in producing the key intermediary metabolites necessary for energy, production through catabolism of amino acids.

Figure 4.8a: AAT activity levels in control and eyestalk ablated *M. malcolmsonii*.

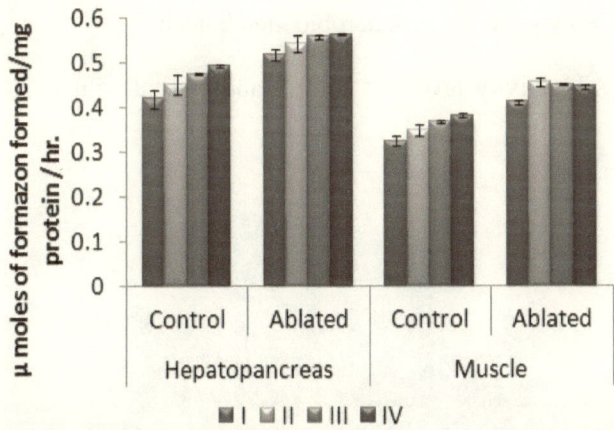

Figure 4.9b: Percent change in AAT activity levels of ablated prawns over controls.

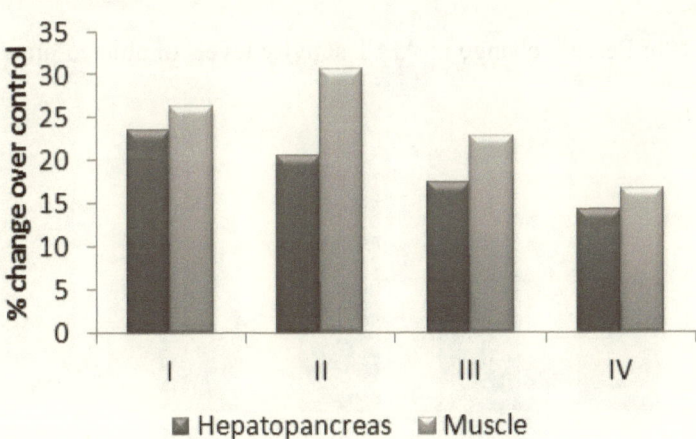

It is also evident from the results that both transaminases are active in the control tissues. Since AlAT promotes the production of pyruvate, it can be taken to represent as a source of energy yielding metabolite and AAT activity indicates the production of oxaloacetate of TCA cycle. Hence higher level of AlAT in muscle represent utilization of FAA in the production of pyruvate, while higher level of AAT in hepatopancreas represent channelling of FAA directly in to TCA cycle thus enhancing energy yield through aerobic metabolism.

Figure 4.10a: AlAT activity levels in control and eyestalk ablated *M. malcolmsonii*.

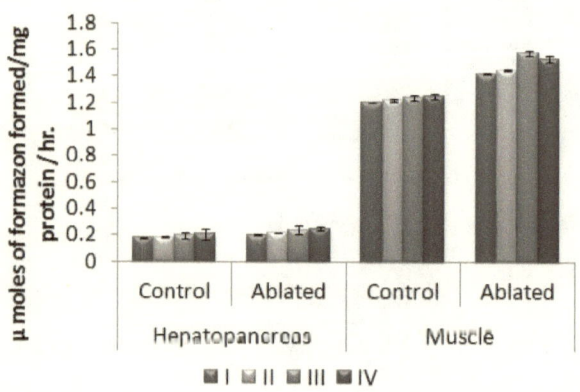

Figure 4.10b: Percent change in AlAT activity levels of ablated prawns over controls.

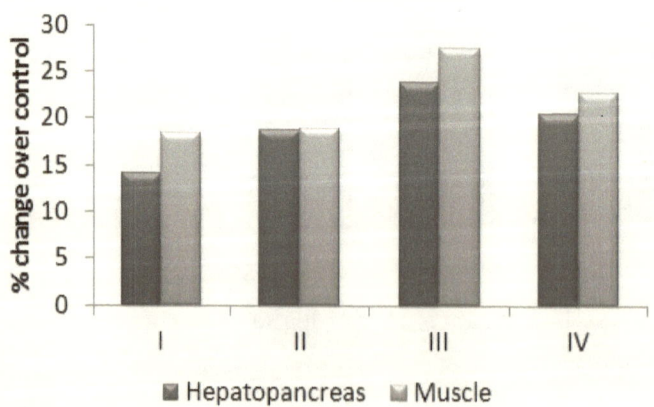

Upon eyestalk ablation there was a significant increase in both AAT and AlAT activities in muscle and hepatopancreas of M. malcolmsonii of all four size groups (Figure 4.9b and 4.10b). Eyestalk ablation is known to cause stress to the animal, resulting in higher metabolic actlvity and thus higher energy requirement. Increased levels of tissue transaminase activity (Figure 4.9b and 4.10b) along with increased oxygen consumption (Figure 3.3a) made it further clear that ablated prawns, meet enhanced energy needs by shunting additional FAA into oxidative pathway. Eyestalk ablation was found to increase respiration rate in crabs (Obuchowicz, 1966; Rangneker and Madhyastha, 1969), prawns (Rosas et al. 1993; Chen and Chia, 1996; Nan et al. 1995; Sierra and Diaz, 1999) and in lobsters (Koshio et al. l992b). This demonstrates that eyestalk ablation which results in the loss of regulatory factors induced additional enerry demand on the prawn. Acceleration in tissue transaminase activities thus may help to provide extra energy required through gluconeogenesis.

In general the crustacean muscle is glycolytic (Hohnke and Scheer, 1970) and therefore exhibits anaerobic metabolism, but in eyestalk ablated tissues there is a tendency to favour aerobic metabolism. Since the free amino acid pool is high in the muscle tissue compared to hepatopancreas, it is likely that the free amino acids may contribute to a greater extent (Through activity of AlAT) towards transamination to step up gluconeogenesis.

Glutamate Dehydrogenase (GDH):

Glutamate dehydrogenase catalyzes the oxidative deamination of glutamate to ammonia and α-ketoglutarate in the presence of NAD (Mayzaud and Conover, 1988). It not only channels nitrogen from glutamate to ammonia but also catalyses the amination of α-ketoglutarate by free ammonia (Harper, 1983).

Results on the activity level of GDH in control and eyestalk abiated *M. malcolmsonii* of four different size groups were presented in figure 4.11a and 4.11b. It is clear from the results that GDH activity of both tissues in control *M. malcolmsonii* increased with an increase in biomass with higher level of activity in hepatopancreas than in the muscle. Braunstein (1939) first suggested the role of glutamate dehydrogenase (GDH) in amino acid metabolism. He proposed that the amino acids are first transaminated with α-ketoglutarate to form glutamate and a corresponding α-keto acid. The newly synthesized glutamate is then oxidized by GDH to release ammonia and regenerate α-ketoglutarate. Since GDH plays a key role in oxidative domination (Harper, 1983), increase in GDH activity corroborates with an increase in ammonia excretion (Figure 3.3a).

Figure 4.11a: GDH activity levels in control and eyestalk ablated *M. malcolmsonii*.

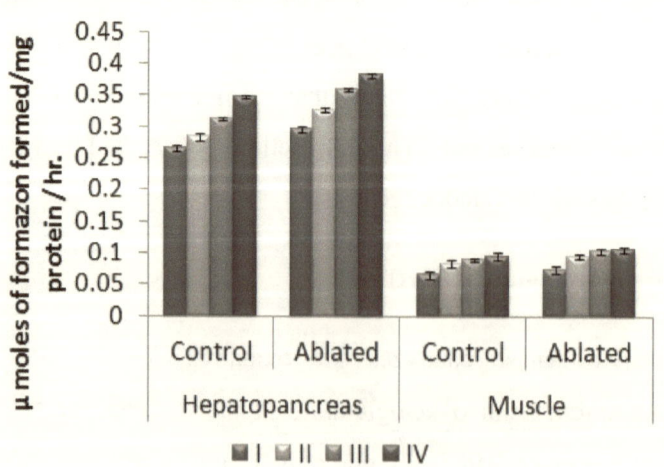

Higher level of GDH activity in hepatopancreas indicates that the ammonia production occurs through the oxidative deamination in this tissue and subscribes to detoxification and mobilization of ammonia and its derivatives.

Figure 4.11b: Percent change in GDH activity levels of ablated prawns over controls.

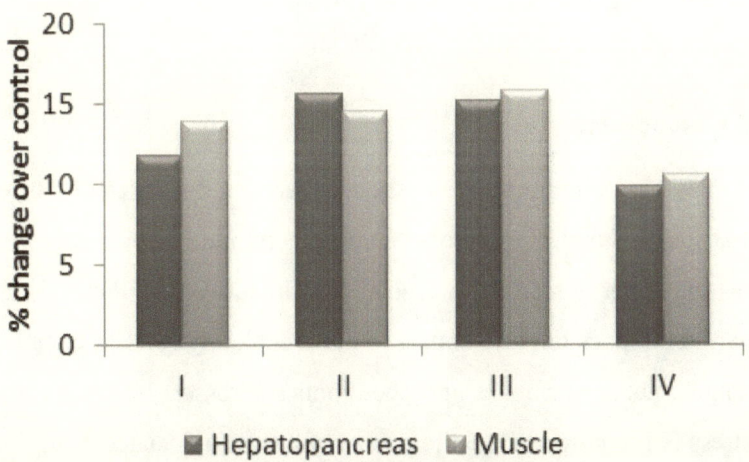

Upon eyestalk ablation, there was a significant increase in GDH activity of both hepatopancreas and muscle in all the four size groups of *M. malcolmsonii* (Figure 4.11b). Higher GDH activity promotes deamination of amino acids produced due to tissue proteolysis. Hence greater elevation in GDH in the ablated HB and muscle of *M. malcolmsonii* indicate higher breakdown of proteins. This is evident through higher levels of protease acticvity upon eyestalk ablation. Thus deamination of some amino acids such as glutamate might have been taken place and incorporated into TCA cycle for the release of necessary energy to meet the demand. Ammonia excretion is also noticed to increase upon eyestalk ablation in all the size groups of M.malcolmsonii (Figure 3.3b). Increased activity of AlAT and AAT

(Figure 4.9a and 4.10a) which lead to increased production of glutamate (Harpaer, et al. 1979) might also be considered for the elevation in GDH and thus arnmonia production in tissues. While Regnault (1979) noticed an elevation in ammonia excretion during early Premoult stage of a sand shrimp, *Crangon crangon* similar effect has been noticed in eyestalk ablated *O. senex senex* by Raghavaiah et al. (1980) suggest an increased transamination and oxidative deamination in the tissues after eyestalk ablation.

Lipid metabolism:

Lipids play very important role in the architectural dynamics of the cell and transport of materials across the cell membrane. Lipids serve not only as vital energy stores and crucial constituents of cellular and sub-cellular membranes but also appear to play regulatory roles as chemical messengers i.e. as hormones and pheromones (Prosser, 1990). As such any stress is found to change the course of events associated with lipid synthesis and membrane structure (Martin et al. 1981). Apart from their structural role, lipids contribute to energy production. Because of high colorific value (9.4 Cal.), lipids infact play a vital role during biochemical adaptation of animals to stress condition (Tayyabha et al. 1981; Swami et al. 1983). Lipids further, contribute towards energy synthesis as an alternative to the carbohydrates (Guyton, 1981; Harper, 1983), act as insulators and as reserve source of energy (Nelson and Cox, 2000). Invertebrates store lipids for reproduction, to regulate buoyancy, or as a response to physical factors such as oxygen levels and temperature (Lawrence, 1976).

Total Lipid (TL) and Triglycerides (Tg):

Figure 4.12a and 4.13a depict results on total lipid and triglycerides in the hepatopancreas, muscle and hemolymph of control and and eyestalk ablated prawns of four different size groups. From results it is evident that total lipid content and triglycerides in the tissues of control *M. malcolmsonii* to increase with increase in biomass. This increase in total lipid and triglycerides in all tissues with increase in biomass suggest that prawns accumulate lipid reserves during growth perhaps for subsequent reproduction. Hill et al. (1992) observed elevated lipid content with dry mass in the benthic amphipods Monoporeia affinis (Lindstrm) and Pontoporeia femorata (Kroyer) suggesting that the animals gain lipids during growth. Thompson and Bergerson (1991) clearly reported that animals store lipids to meet the energetic demands of reproduction and the hatchery reared Colorado squawfish was found to increase the lipid content with increase in total length.

Figure 4.12a: Total lipid content in control and eyestalk ablated *M. malcolmsonii*.

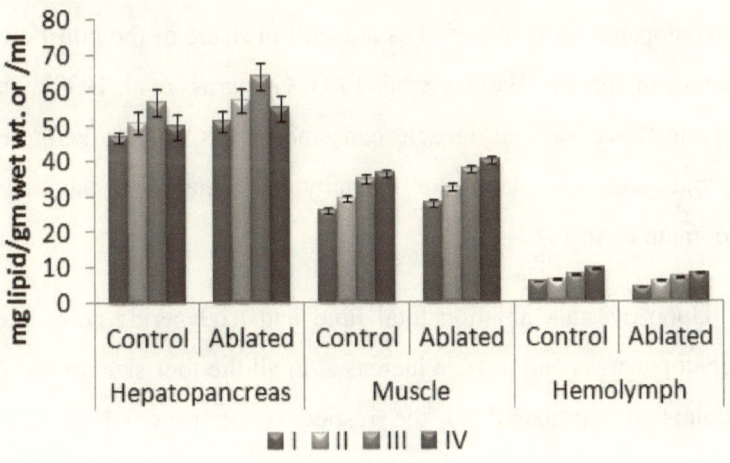

Figure 4.12b: Percent change in total lipid content of ablated prawns over controls.

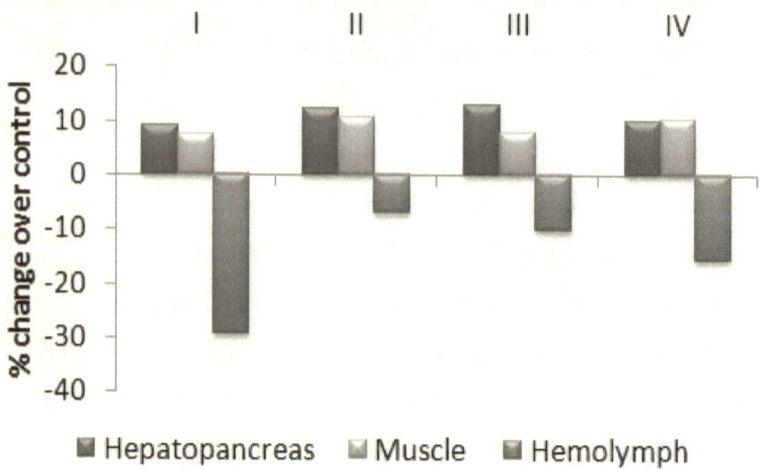

It is also evident from the results that the total lipid and triglycerides were high in the hepatopancreas than in muscle and hemolymph in control prawns, suggesting the significance of hepatopancreas as storage site (Chang and O' Connor, 1983; Muriana et al. 1993). Hepatopancreas of crustaceans is well known as rich source of lipids (Erribabu and Hanumantha Rao, 1983) and thus lipid content of the hepatopancreas is regarded as a useful measure of the nutritional condition of the prawn (Barclay et al. 1983; Glencross et al. 1999). The non hepatic tissue such as muscle can store lipids only to a limited extent, but seems to lack the capacity to metabolise the same (Surendranath et al. 1987).

Upon eyestalk ablation total lipid and triglyceride content of both hepatopancreas and muscle increased in all the four size groups of M.malcolmsonii compared to the respective controls whereas the hemolymph total lipid and triglyceride content decreased (Figure 4.12b and 4.13b). This suggests that there must be either synthesis or accumulation of certain lipid fractions due to insuffiiciency of eyestalk hormones which otherwise would regulate lipid accuulation.

Figure 4.13a: Triglyceride levels in control and eyestalk ablated *M. malcolmsonii*.

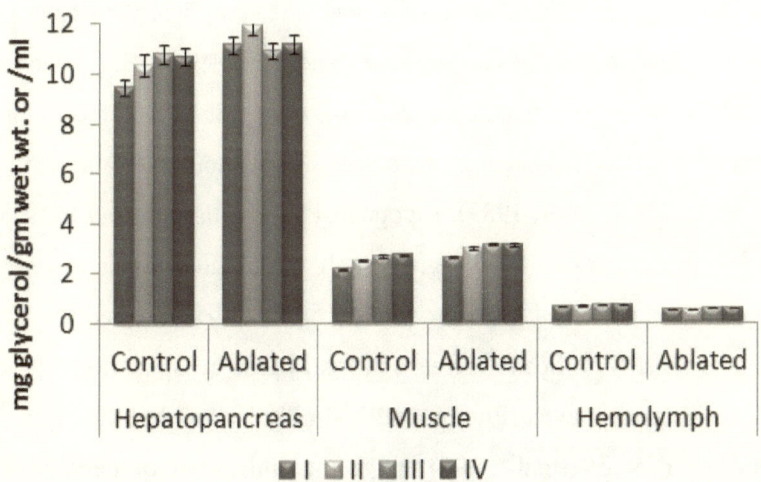

Figure 4.13b: Percent change in triglyceride levels of ablated prawns over controls.

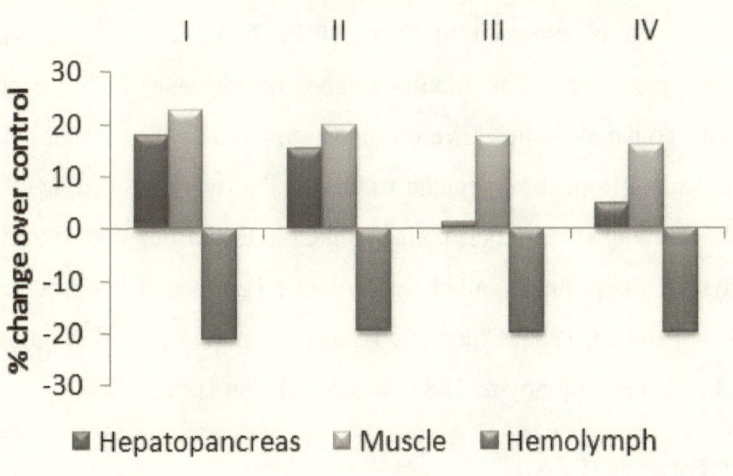

Reports show that hormonal factors from eyestalk inhibit lipid synthesis in the hepatopancreas of crustaceans (O' Connor and Gilbert, 1968, 1969), while some reports demonstrate eyestalk ablation to significantly increase tissue lipid content. Eyestalk ablation in the crab, Pachygrapsus carassipes (Randall) and the crayfish, Procambarus increased lipid biosynthesis (Highnam and Hill, 1977). Similar trend was obtained in the hepatopancreas of two crabs, Gecarcinus and Orconectus uirilis incubated *in vitro* with and without eyestalk extracts (Chang and O' Connor, 1983). Lipogenesis in the tissues after eyestalk ablation in the present study agrees well with higher levels of Krebs cycle enzyme activity for the production of NADPH, required for fatty acid synthesis. Significant decrease in total lipids in the hemolymph of eyestalk ablated prawns (Figure 4.12b) similar to the observation in C. granulates (Santos et al. 1997) suggest mobilization of hemolyrnph lipids for deposition in other storage structures.

A close link between hemolymph, ovary and extra ovarian synthetic sites is expected to throw more light on the underlying interrelationship between moulting and rproductron in crustacea. Although gonadial tissues have not been explored in the present study lipid transport from the lipogenic tissues to the ovary as well as other organs including epidermis is mainly linked in the earlier studies to two hemolymph lipoproteins called lipoprotein I (Lp I) and lipoprotein II (Lp II) (Chapman, 1980; Chino and Kitazawa, 1981; Lee and Puppione, 1988; Laverdure and Soyez, 1988; Jugan and Van Herp, 1989).

Free fatty acids (FFA):

Figure 4.14a summarize results on free fatty acid content of the hepatopancreas, muscle and hemolymph in control and eyestalk ablated

M. rnalcolmsonii of four different size groups. It is clear from the results that the tissues free fatty acid content in both hepatopancreas and muscle increased with increase in biomass (Figure 4.14a). Free fatty acids arise from the break down of triacylglycerols, by the action of lipase, which are resynthesised in tissues to acyl co-A and re-esterified with glycerol 3-PO_4 to form triacylglycerol (Harper, 1983). Thus a continuous synthesis and break down of free fatty acids occur in tissues. It is also evident that free fatty acid content is more in hepatopancreas than in muscle and hemolymph of control prawns suggesting hepatopancreas to be the main tre for synthesis and metabolism of free fatty acids.

Figure 4.14a: Free fatty acid levels in control and eyestalk ablated *M. malcolmsonii*.

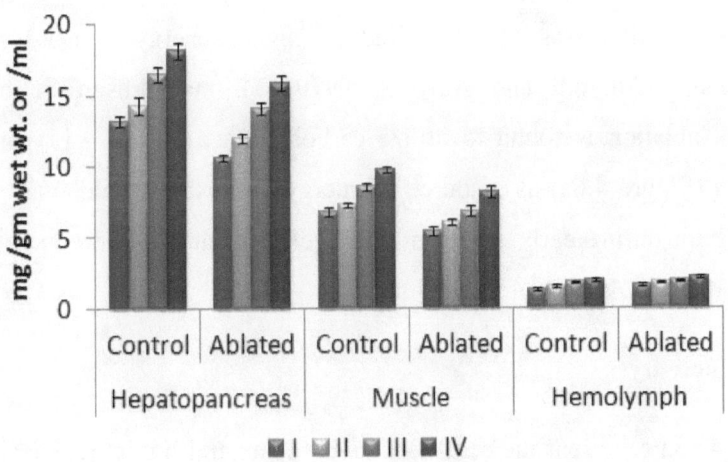

Upon eyestalk ablation FFA content of both hepatopancreas and muscle decreased in all four size groups of M. malcolmsonii compared to the respective controls (Figure 4.14b) indicating synthesis of lipid by hepatopancreas and muscle tissue.

Figure 4.14b: Percent change in free fatty acid levels of ablated prawns over controls.

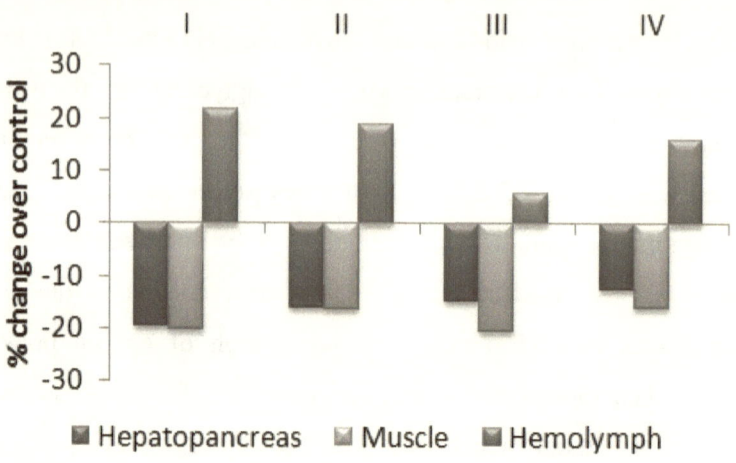

Higher levels of FFA in hemolymph upon eyestalk ablation further indicate their transportation to the sites of lipogenesis. Although eyestalk ablation was found to induce higher energy demand in crustaceans (Adiyodi and Adiyodi, 1970) M. malcolmsonii upon eyestalk ablation is found to utilize carbohydrates (Figure 4.1a) and proteins (Figure 4.6a) as a source of energy and exhibit lipid sparing action both during early stages of life cycle and during reproductive phase (Figure 4.12a and 4.13a).

Lipase activity:

Figure 4.15a represent the results on assay of neutral lipase activity in the control and eyestalk ablated prawns, M. malcolmsonii. It is evident from the present results that the lipase activity is more in hepatopancreas of all size groups than muscle tissue. Hepatopancreas being the main metabolic centre and with high total lipid content is expected to have higher lipase activity. When lipids are ingested, lipase in the digestive tract mediates hydrolysis of total lipids to free fatty acids and glycerol (Kleine, 1967; Gilbert, 1967). Lipase activity in muscle is due to the presence of lipid fractions like glycolipids, phospholipids and lipoproteins which are more associated with structural formation, rather than metabolic activities.

From the results it is also evident that lipase activity is gradually increased in both tissues of control and ablated prawns, with increase in biomass (Figure 4.15a). Increased total lipid and triglycetrides (Figure 4.12a and 4.13a) despite a parallel increase in lipase activity with increase in size (Figure 4.15a) clearly suggest higher rate of accumulation of lipids than utilization with progress in biomass. However decreased lipase activity of both hepatopancreas and muscle in four size groups of *M. malcolmsonii* upon eyestalk ablation compared to the respective controls (Figure 4.15b) was noted.

Figure 4.15a: Lipase activity levels in control and eyestalk ablated *M. malcolmsonii*.

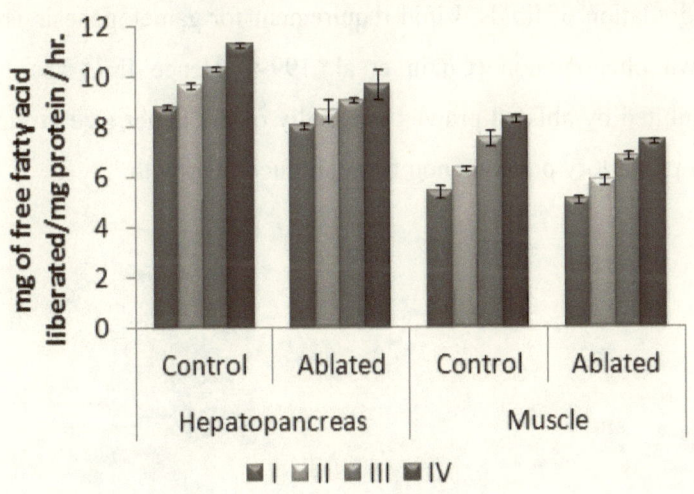

Figure 4.15b: Percent change in Lipase activity levels of ablated prawns over controls.

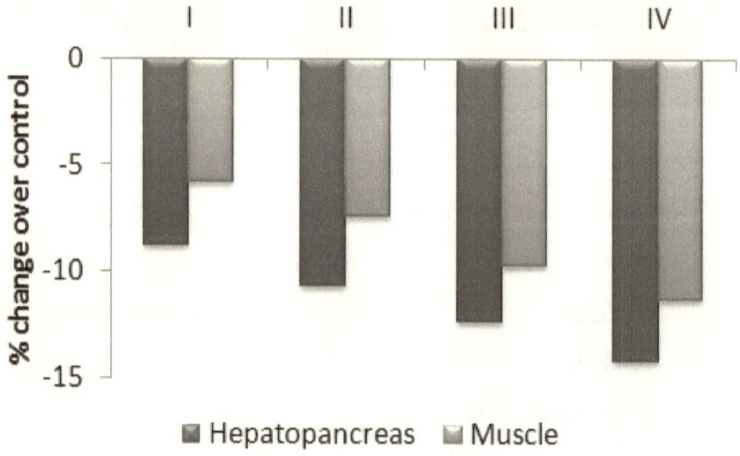

The reduction in lipase activity is might be due to reduction in eyestalk hormones especially like gonadal inhibiting hormone which lead to the promotion of reproductive growth through elevation in gonad stimulating hormone (Wilder et al. 1994), prawns may prefer to reduce degradation of lipids. Lipid requirement for gametogenesis is a well known phenomenon (Cuzin et al. 1999). Hence lipid sparing action exhibited by ablated prawns especially of the larger size groups suggests a prepartory phenomenon for reproductive growth.

Summary and Conclusions

Among the freshwater prawns *M. malcolmsonii* (H. Milne Edwards) and *M. rosenbergii* (De Man) formed the cynosure of aquaculturists. Ever since indiscriminate commercial exploitation started, the number of the genus *Macrobrachium* started depleting to a very great extent. Thus, there is an urgent need to overcome this problem through enriching the production of this species. The best way to do this is to understand the physiology and biochemistry of this species. Eyestalk ablation, the practice followed in hatcheries is known to cause precocious moulting and reproduction, which are energy demanding processes. The pattern of energy allocation affects the survival and fitness of the organisms. Characterization of biomass specific moulting stages, apportioning of obtained energy to different energy demanding processes (bioenergetics) and associated changes in metabolite and enzyme activities of nutrient sources in eyestalk ablated prawns of four different size groups formed the focal theme of this thesis. The freshwater prawn, *Macrobrachium malcolmsonii* was used as an experimental animal. Prawns were acclimated to laboratory conditions and fed *ad libitum* with commercial feed daily. Influence of unilateral eyestalk ablation on moulting, energy acquisition, allocation and metabolism has been studied in comparison to normal (control) individuals.

Investigations on the biology and morphometry of the species made in Chapter- I highlight its positional suitability for culture under captivity. The regression analysis carried out in the present study confirms that all morphological characters observed account for positive allometry and indicate regular and positive growth pattern of

M. malcolmsonii further suggesting that growth in this species is rapidly attained over a restricted series of moults.

A detailed description of moult cycle stages in *M. malcolmsonii* given in Chapter-II provided scope to identify all the four major moult cycle stages and respective sub stages (in parenthesis) viz., Postmoult (A_1, A_2, and B), intermoult (C), premoult (D_0, D_1, D_2, D_3 and D_4) and ecdysis (E). The development of setae of the uropod clearly provided evidence to identify substages of the premoult stage. Premoult stage took maximum time duration followed by intermoult stage. Post moult stage is a brief phase suggesting the quick hardening of the newly formed cuticle in *M.malcolmsonii*.

Moult cycle duration in both control and ablated prawns increased significantly with increase in biomass indicating size dependency of moulting duration in *M. malcolmsonii*. In the total moult duration of control prawns premoult, postmoult and intermoult were found to comprise 73.68%, 8.77% and 17.55% respectively. Eyestalk ablation significantly lowered the intermoult (47.21%), premoult (9.98%) and postmoult (13.97%) duration.

Energy consumption increased with increase in biomass in both control and eyestalk ablated prawns which indicate a positive correlation between biomass and feeding capacity. Eyestalk ablation was found to induce increased food ingestion through enhanced gastric evacuation caused due to higher energy demands of moulting and reproduction.

To know the variations in the energy loss in *M. malcolmsonii* in relation to size and eyestalk ablation variables such as faeces (egestion), absorbed energy and absorption efficiency; ammonia excretion (Excretion), assimilated energy and assimilation efficiency; routine respiratory metabolism (R_{rout}) and apparent specific dynamic action (ASDA) were studied. All variables exhibited increase with increase in biomass. Increased energy loss in faeces upon eyestalk ablation in each size group was found to be due to lack of increase in digestion and absorption abilities on par with food ingestion. Higher energy demands upon eyestalk ablation were found to result in increased catabolism and utilization of proteins and higher energy loss in excretion (ammonia).

Routine metabolism increased with increase in biomass showing positive correlation to the absolute increase in active tissue and respiratory ability with biomass. Eyestalk ablation was found to increase routine metabolism indicating increased metabolic costs of maintenance associated with tissue repair and anabolic activity. Decreased percentage of energy allocation to routine metabolism in size group IV along with increased amount of energy potentially available for growth exhibit the trade - off mechanism present in *M. malcolmsonii* which facilitate increased energy allocation to reproduction and thus to enhanced fecundity.

Increased ASDA with increase in biomass in *M. malcolmsonii* was found to be the direct effect of the amount of food ingested. Higher ASDA upon eyestalk ablation indicated increased costs of food processing which was found to be associated with increased food ingestion.

Increased energy potentially available for growth (scope for growth : SFG) along with increased biomass was found to be an effect of absolute enhancement in food ingestion. But the same upon eyestalk ablation was found to be the combined effect of enhanced food ingestion and assimilation efficiency.

Changes in energy allocation to different components of energy budget like faeces and ammonia excretion (wastage), maintenance and ASDA (respiration) and scope for growth (SFG) with increase in biomass and upon eyestalk ablation suggest variations in digestion ability, gut capacity, absorption, assimilation, costs of maintenance and efficiency of growth respectively.

Increased proportions of consumed energy loss in wastage with increase biomass showed that although energy ingestion increases with biomass due to lack of equivalent increase in efficiencies of digestion, absorption and assimilation, prawns tend to lose increased proportion of consumed energy through wastage. Percent energy allocation to respiration decreases although maintenance costs increase with increase in biomass perhaps due to decreased duration of activity as biomass increases.

Increased total carbohydrate, glycogen and hemolymph glucose in both muscle and hepatopancreas with increase in biomass indicates the ability of *M.malcolmsonii* to synthesize and store energy reserves for growth.

But decreased total carbohydrate and glycogen contents of the muscle and hepatopancreas and increased hemolymph glucose levels after eyestalk ablation show possible utilization of total carbohydrate and glycogen to overcome the consequences of eyestalk ablation and elevated levels of readily utilizable glucose in hemolymph through breakdown of glycogen to meet the energetic demands.

Enhanced levels of lactate dehydrogenase with increase in biomass and with eyestalk ablation along with increased levels of oxidative enzymes such as succinate dehydrogenase and malate dehydrogenase which facilitate oxidative phosphorylation and thus ATP generation indicate increased rate of conversion of lactate to pyruvate. Higher pyruvate concentrations thus ensure energy supply through the oxidative metabolism.

Increased total protein content with a parallel increase in free amino acids and protease activity both with increase in biomass and upon eyestalk ablation clearly demonstrate net accumulation of proteins in hepatopancreas and muscle despite increased breakdown of protein for energy production.

Increased glutamate dehydrogenase activity with increase in biomass and upon eyestalk ablation further indicate active deamination of amino acids produced due to tissue proteolysis, the effect which has been clearly reflected in the elevation of excretory loss. Elevation in AAT and AlAT with increase in biomass and eyestalk ablation demonstrates reorganization of amino acids for the synthesis of necessary proteins, the effect which has been noticed in enhancement of total protein content.

Total lipid and triglyceride contents increased with increase in biomass in hepatopancreas and muscle with simultaneous increase in lipase activity and enhanced production of free fatty acids demonstrating net accumulation of lipids during growth for subsequent use despite enhanced utilization of lipids, which are high energy storage substances to meet routine energy demands. On the other hand upon eyestalk ablation total lipids and triglycerides increased in hepatopancreas and muscle with simultaneous decrease in lipase activity, free fatty acids and hemolymph lipids clearly indicate the lipid sparing action exhibited by *M. malcolmsonii* perhaps to promote reproductive growth. This can be considered an effect of decrement in gonad inhibiting hormone (GIH) caused through eyestalk ablation.

This study clearly demonstrated the effects of size and eyestalk ablation on moult duration / staging, pattern of energy allocation for growth and the metabolic alterations underlying energy allocation in *M.malcolmsonii* for the first time and thus, might contribute, in a modest way, to overall understanding of the energy dynamics of growth in this species.

References:

Abramowitz, A.A., Hisaw, F.Z. and Papandrea, D.V. 1944. The occurrence of diabetogenic factor in the eyestalks of crustaceans. Biol. Bull., 86: 1- 4.

Abramowitz, R.K. and Abramowitz, A.A. 1940. Moulting, growth and survival after eyestalk removal in *Uca pugilator*. Biol. Bull., 78 : 179 - 188.

Adams, S.M., McLean, R.B. and Parrotta, J.A. 1982. Energy partitioning in large mouth bass under conditions of seasonally fluctuating prey availability. Transactions of the American Fisheries Society, 111: 549 - 558.

Adiyodi, R.G. 1985. Reproduction and its control. In: The Biology of Crustacea, (eds. Bliss, D.E. and L.H. Mantel), Vol. 9, pp. 147 - 216. Academic Press, Orlando, Fh, (USA).

Adiyodi, K.G. and Adiyodi, R.G. 1970. Endocrine control of reproduction in decapod crustacea. Biol. Rev., 45: 121 - 165.

Adiyodi, R.G. 1968. On reproduction and moulting in the crab, *Paratelphusa hydrodromous*. Physiol. Zool., 41 : 204 - 209.

Ahmed, S.K.J. 1984. Fishery of freshwater prawn *Macrobrachium malcolmsonii* in Sambalpur district and Prospects of its culture in western Orissa. In: Souvenir of the seminar on Freshwater fisheries and Rural Development, Rourkela, Orissa, 6-7 April, Section II : 1-7.

Aiken, D.E. and Waddy, S.L. 1976. Controlling growth and reproduction in the American lobster. Proc. Annu. Meet. World Maricult. Soc., 7 : 415 - 430.

Alikunhi, K.H., Poernomo, A., Adisukresno, S., Budiano, M. and Busman, S. 1975. Preliminary observations on induction of maturity and spawning in *Penaeus monodon* Fabricius and *Penaeus merguiensis* de Man by eyestalk extirpation. Bull. Shrimp. Cult. Res. Cent., 1 : 1 - 11.

Allen, J.R.M. and R.J. Wootton. 1983. Rate of food consumption in a population of threespine sticklebacks, *Gasterosteus aculeatus*, estimated from the faecal production. Envir. Biol. Fish., 8 : 157 - 162.

Allen, J.A. 1962. Observations on Spirantocaris from Northumberland waters. Crustaceans, 3 : 227 - 238.

Ananthakrishnan, K.V., Katre, S. and Reddy, S.R. 1981. Influence of endogenous factors on the pattern of ammonia excretion in the prawn *Macrobrachium lanchesteri* (de Man). Indian J. Exp. Biol., 19 : 42 - 45.

Anilkumar, G. and Adiyodi, K.G. 1985. The role of eyestalk hormones in vitellogenesis during the breeding season in the crab, *Paratelphusa hydrodromous* (Herbst). Biol. Bull., 169 : 689 - 695.

Anne Bond, J., C.R.M. Gonzalez and B.P. Bradley. 1993. Age - dependent expression of proteins in the cladoceran *Daphnia magna* under normal and heat ☐ stress conditions.

Anonymous, 1993. BOBP. Manual on freshwater prawn seed production. BOBP, Madras, 31pp.

AOAC. 1981. Official methods of analysis, 12th edition. Association of Official Analytical Chemists. Washington D.C. 1094.

Arnstein, D.R. and Beard, T.W. 1975. Induced maturation of the prawn *Penaeus orientalis* Kishinouye in the laboratory by means of eye stalk removal. Aquaculture, 5 : 411 - 412.

Assem, H. and W. Hanke. 1983. The significance of the amino acids during osmotic adjustment in teleost fish. I. Changes in the euryhaline *Sarotherodon mossambicus*. Comp. Biochem. Physiol., 74A: 531-536.

Bano, Y., S.A. Ali and H. Tariq.1981. Effect of sublethal concentrations of DDT on muscle constituents on an air breathing cat fish *Clarias batrachus*. Proc. Indian Acad. Sci., (Ani.Sci)., 90:33-37.

Barclay, M.C., W. Dall and D.M.Smith. 1983. Changes in lipid and protein during starvation and moulting cycle in tiger prawn, *Penaeus esculentus*. J. Exp. Mar. Biol. Ecol., 68:229-244.

Bartell, S.M., J.E. Breck, R.H. Gardner and A.L Brenkert. 1986. Individual parameter perturbation and error analysis of fish bioenergetic models. Canadian Journal of fisheries and Aquatic Sciences, 43 : 160 - 168.

Bauchau, A.G., Mengeot, J.C. and Oliver, M.W. 1968. Action de la seratonin et de la hormone diabetogene des crustaces sur la phosphorylase musculature. Gen. Comp. Endorcinol., 2 : 132 - 138.

Beamish, F.H.W. 1974. "Apparent specific dynamic action of large mouth bass, *Micropterus salmoides"*, J. Fish. Res. Bd. Can., 31 : 1763 - 1769.

Beamish, F.W.H and E.A. Trippel. 1990. Heat increment: A static or Dynamic Dimension in Bioenergetic Models? Transactions of the American Fisheries Society, 119 : 649 - 661.

Beauchamp, D.A., D.J. Stewart and G.L. Thomes. 1989. Corroboration of a bioenergetics model for sockeye salmon. Transactions of the American Fisheries Society, 118 : 597 - 607.

Bellany, D. 1962. The endogenous citric acid cyel intermediates and amino acids of mitochondria. Biochem. J., 82 : 218 - 224.

Bellonci, G. 1882. 'Nuove ricerche sulla struttura del ganglio ottico dell *Squilla mantis'*. Memorie della reale Accademia delle Scienze dell; Istituto di Bologna, 3 : 419 - 426.

Benke, A.1996. Secondary production of Macroinvertebrates, p.557 - 578. In Hauer, F.R., and G.A.Lamberti (eds). Methods in stream ecology : Academic, New York.

Bhimachar, B.S. 1962. Information on prawns from Indian waters- synopsis of biological data. Proc. Indo-pacific Fish. Coun., 10 : 124 - 133.

Bliss, D.E. 1968. Transition from water to land in decapod crustaceans. Am. Zool., 8 : 355-392.

Bliss, D.W. 1953. Endocrine control of Metabolism in the land crab, *Gecarcinus lateralis* (Freminville) I. Differeces in the respiratory metabolism of sinus glandless and eyestalkless crabs. Biol. Bull., 104: 275 - 296.

Bohlken, S. and J. Joosse. 1982. The effect of photoperiod on female reproductive activicty and growth of the freshwater pulmonate snail *Lymnaea stagnalis* kept under laboratory breeding conditions. International journal of invertebrate reproduction, 4 : 213 - 222.

Boisclair, D. and Legget, W.C. 1988. An experimental evaluation of the Elliott, Person and Eggers models for estimating fish daily ration. Canadian Journal of Fisheries and Aquatic Sciences, 45 : 138 -145.

Bollenbacher, W.E., Broast, D.W. and O☐Connor, J.D. 1972. Endocrine regulation of lipid synthesis in decapod crustaceans. Am. Zool., 12 : 381 - 384.

Bonner, J.J. 1982. An assessment of the ecdysteroid receptor of *Drosophila*. Cell, 30 : 7 - 8.

Borsook, H. 1936. The specific dynamic action of protein and amino acids in animals. Biol. Rev. 11 : 147.

Brafield, A.E. and M.J. Llewellyn. 1982. Animal Energetics. Glasgow, Blackie.

Brafield, A.E. 1985. Laboratory studies of energy budgets. Pages 257 - 281 in P. Tytler and P. Calow (editors). Fish energetics, new perspectives. The John Hopkins University press, Baltimore.

Braunstein, A.E. 1939. The enzyme system of transamination its mode of action and biological significance. 143 : 609-610.

Bray, W.A. and Lawrence, A.L. 1992. Reproduction of *Penaeus* species in captivity, pp. 93-170. *In*. A. Fast and L.J. Lester (eds.), Culture of Marine Shrimp : Principles and Practices, Elsevier Sci. Publ., Amsterdam.

Brett, J.R. and Zala, C.A. 1975. Daily pattern of nitrogen excretion and oxygen consumption of sockeye salmon (*Oncorhynchus nerka*) under controlled conditions. J. Fish. Res. Board Can., 32 : 2479 - 2486.

Brett, J.R. 1976. Feeding metabolic rates of young sockeye salmon, *Oncorhynchus nerka* in relation to ration level and temperature. Fish Mar. Serv. Res. Dev. Tech. Rep., 675 : 43.

Brett, J.R. and Grooves, T.D.D. 1979. Physiological Energetics. In Fish physiology, vol. VIII, (eds W.S. Hoar, D.J. randall and J.R. Brett), pp. 280 - 352. New York: Academic press.

Brett, J.R. 1972. The metabolic demand for oxygen in fish, particularly salmonids, and a comparison with other vertebrates. Respir. Physiol., 14 : 151 - 170.

Brett, J.R. 1971. Energetic response of salmon to temperature. A study of some thermal relations in the physiology and freshwater ecology of sockeye salmon (*Oncorhynchus nerka*). Am. Zool., 11 : 99 - 113.

Brett, J.R. 1964. The respiratory metabolism and swimming performance of young sockeye salmon. J. Fish. Res. Board Can., 21 : 1183 - 1226.

Browdy, C.L. and Samocha, T.M. 1985. The effect of eyestalk ablation on spawning, moulting and mating of *Penaeus semisulcatus* de Haan. Aquaculture, 49 : 19 - 29.

Brown, F.A.Jr. and Cunningham, O. 1939. Influence of sinus gland of crustacean on normal viability and ecdysis. Biol. Bull., 77 : 104 - 114.

Brown, J.H. 1991. Freshwater prawns. In ; Production of Aquatic Animals : Crustaceans, Molluses, Amphibians and Reptiles, C.E. Nash (ed) pp 31-43. Elsevier Science Publication. B.V., Amsterdam.

Brown, Jr. A., Mc Vey, J., Middleditch, G.S., Lawrence, A.L., Scott, B.M. and Williams, T.D. 1980. Maturation and spawning of *Penaeus stylirostris* under controlled laboratory conditions. Proc. Annu. Meet. World Maricult. Soc., 11 : 488 - 499.

Bruce, M.J. and Chang, E.S. 1984. Demonstration of a moult inhibiting hormone from the sinus gland of the lobster, *Homarus americanus*. Camp. Biochem. Physiol., 79 A : 421 - 424.

Buttery, P.J. and Annison, E.F. 1973. Considerations of the efficiency of amino acid and protein metabolism in animals. Pages 141-171 in J.G.W. Jones, editor. The biological efficiency of protein production. Cambridge University Press, London.

Caillouet, C.W. Jr. 1973. Ovarian maturation by eye stalk ablation in pink shrimp *Penaeus duorarum* Burkenroad. Proc. Annu Workshop world Maricult. Soc., 3 : 205 - 225.

Calow, P. 1981a. Resource utilization in reproduction. In C.R. Townsend and P. Calow (Eds), Physiological Ecology. An evolutionary approach to resource use : 245 - 270. Oxford : Blackwell scientific publications.

Calow, P. 1981b. Growth in lower invertebrates. In M. Rechcigl (Ed), Comparative Animal Nutrition, 4 : 53 - 76. Basel: S. Karger.

Calow, P., Beveridge, M. and Sibly,R. 1979. Heads and Tails: Adaptational aspects asexual reproduction in freshwater triclads. American zoologist, 19 : 715 - 728.

Calow, P. 1984. Exploring the adaptive landscapes of invertebrate life cycles. Advances in invertebrate reproduction, 3: 329 - 342.

Calow, P. 1977. Conversion efficiencies in heterotropic organisms. Biol. Rev, 52 : 385 - 409.

Camien, M.N., Sarlet, H., Duchateau, G. and Florkin, M. 1951. Nonprotein amino acids in muscle and blood of some marine and freshwater crustaceans. J. Biol. Chem., 193, 881 - 885.

Campana, S.E. and Neilson, J.D. 1985 : Microstructure of fish otoliths. Canadian Journal of Fisheries and Aquatic Sciences, 42 : 1014 - 1032.

Carefoot, T.H. 1990. Specific dynamic action (SDA) in the supralittoral isopod, *Ligia pallasii* : Identification of components of apparent SDA and effects of dietary amino acid quality and content on SDA.

Carmichael, G.J., Tomasso, J.R., Simco, B.A. and Davis, K.B. 1984. Characterization and alleviation of stress associated with hauling largemouth bass. Transactions of the American Fisheries Society, 113: 778-785.

Carrol, N.V. Longley, R.W. and Raw, J.H. 1956. Glycogen determination in liver and muscle by use of Anthrone reagent, J. Biol. Chem, 22, 583.

Carter, C. G. and Brafield, A.E. 1992 : The relationship between specific dynamic action and growth ingrass carp, *Ctenopharyngodon idella* (Val). Journal of Fish Biology, 40 : 895 - 907.

Carter, C.G. and Brafield, A.E. 1991 ; The bioenergetics of grass carp, *Ctenopharyngodon idella* (Val) : Energy allocation at different planes of nutrition. Journal of Fish Biology, 39 : 873-887.

Castro, H., Battaglia, J. and Virtanen, E. 1998. Effects of FinnStim on growth and sea water adaptation of Coho salmon Aquaculture, 168 (1 - 4) : 423 - 429.

Caulton M.S. 1978. The importance of hebitat temperatures for growth in the tropical cichlid *Tilapia rendalli* Boulenger. J. Fish.Biol., 13: 99-112.

Chakravarty, M.N. 1992. Effect of eystalk ablation on moulting and growth in prawn *Macrobrachium rosenbergii*. Indian Journal of Marine Sciences, 21 : 287 - 289.

Chandrasekharan. V.S. and Sharma, A.P. 1997. Biology and culture of freshwater prawns in North India. Fishing Chimes, 16 (11) : 7 - 9.

Chang, E.S. 1985 Hormonal control of moulting in decapod crustacea. Amer. Zool., 25: 179-185.

Chang, E.S., 1995. Physiological and biochemical changes during the molt cycle in decapod crustaceans : an overview. J. Exp. Mar. Biol. Ecol., 193 : 1 - 14.

Chang, E.S. 1997. Chemistry of crustacean hormones that regulate growth and reproduction. In "Recent advances in marine biotechnology", M. Fingerman, R. Nagabhushanam, and M.F. Thompson, eds., pp 163-178. Oxford and IBH Publishing, Co., New Delhi.

Chang, E.S., 1993. Comparative endocrinology of molting and reproduction : Insects and crustaceans. Annu. Rev. Entomol., 38 : 161-180.

Chang, E.S., and O, Connor, J.D. 1983. Metabolism and transport of carbohydrates and lipids. In : The Biology of crustacea. Vol. V. Internal Anatomy and Physiological regulation. Edited by L.H. Mantel. Academic Press. New York. London. pp. 263 - 287.

Chaplin, A.E., Huggins, A.K., and Munday, K.A. 1967. The distribution of 1-α aminotransferases in *Carcinus maenas*. Comp. Biochem. Physiol., 20: 195 - 198.

Chapman, M.J. 1980. Animal lipoproteins : Chemistry, Structure and Comparative aspects. J. Lipid Res., 21 : 789-853.

Charmantier - Daures, M., Charmantier, G., Van Deijnen, J.E., Van Herp, F., thuet, P., Trilles, J. - P. and Aiken, D.E. 1988. Isolement d□um facteur pedonculaire intervenant dans le controle neuroendocrine du metabolisme hydromineral de *Homarus americanus* (crustaces, Decapoda). Premiers resultats. C.R. Acad. Sci. Paris, 307 : 439 - 444.

Charmantier-Daures, M. and De Reggi, M. 1980. Aspects preliminaires des variations hemolymphatiques du taux, decdysteroides chez *Pachygrapsus marmoratus* (Crustace, Decapode) : Influence de la regeneration intensive et de lablation des organes Y. Bull. Soc. Zool. France, 105 : 81 - 86.

Charmatier-Daures, M. and Vernet, G. 1974. Nouvelles donnees sur le role de lorgane Y dans le deroulement de la mue chez Pachygrapsus marmoratus (Decapode, Grapside). Influence de la regeneration intensive. C.R. Acad. Sc. Paris, 278 : 3367 - 3370.

Charniaux-Cotton, H. 1985. Vitellogenesis and its control in Malacostracan crustacea. Amer. Zool., 25 : 197 - 206.

Chaves, A.R. 2001. Effects of sinus gland extracts on mandibular organsize and methyl farnesoate synthesis in the crayfish. Comp. Biochem. Physiol. 128 (A) : 327 - 333.

Chaves, A.R. 2000. Effect of X-organ sinusgland extract on [35 S] methionine incorporation to the ovary of th ered swamp crayfish *Procambarus clarkii*. Camp. Biochem. Physiol., 126A : 407 - 413.

Chen J.C., Nan,F.H. and Kuo, C.M. 1991. Oxygen consumption and ammonia-N excretion of prawns *(Penaeus chinensis)* exposed to ambient ammonia. Arch. Environ. Contam. Toxicol., 21: 377-382.

Chen, C.H., and Lehninger, A.I. 1973. Respiration and phosphorylation by mitochondria from the hepatopancreas of the blue crab, *Callinectes sepidus*. Arch. Biochem. Biophys., 154 - 445.

Chen J.C. and Y.Z. Kou. 1991. Accumulation of ammonia in the haemolymph of *Penaeus japonicus* exposed to ambient ammonia. Dis. Aquat. Org., 11 : 187 - 191.

Chen, J.C. and S.H. Lai. 1992. Oxygen consumption and ammonia excretion of *Penaeus japonicus* adolescents exposed to ambient ammonia. Comp. Biochem. Physiol., 102 (1): 129-133.

Chen, S., Wu, J., Huner, J.V. and Malone, R.F. 1995. Effects of temperature upon ablation-to-molt interval and mortality of red swamp crawfish (*Procambarus clarkii*) subjected to bilateral eyestalk ablation. Aquaculture. 138 (1-4): 191 - 204.

Chen, J.C. and Chia, P.G. 1996. Effects of unilateral eyestalk ablation on oxygen consumption and ammonia excretion of juvenile *Penaeus japonicus* Bate at different salinity levels. J. Crust. Biol., 15 (3) : 434-443.

Chino, H. and Kitazawa, K. 1981. Diacyglycerolcarrying lipoprotein of hemolymph of the locust and some insects, J. Lipid Res., 22 : 1042-1052.

Cho, C.Y., Slinger, S.J. and Bayley, H.S. 1982. Bioenergetics of salmonid fishes : energy intake, expediture and productivity. Comp. Bioch. physiol., 73B : 25 - 41.

Cho, C.Y., Bayley,H.S. and Slinger, S.J. 1976. Energy metabolism in growing rainbow trout : partition of dietary energy in high protein and high fat diets. Pages 299-302 in M. Vermorel, editor.

Proceedings of the seventh symposium on energy metabolism. Vichy, France. Bussac, Clermont-Ferrand, France.

Claybrook, D.L. 1983. Nitrogen metabolism. In : The Biology of crustacea. Vol. 5., Ed. Mantel, L.H. Academic Press, New York, London, pp. 163 - 213.

Clifford, H.C. III and R.W.Brick. 1978. Protein utilization in the freshwater shrimp *Macrobrachium rosenbergii*. Proc. 9th Ann-Meeting World Maricult. Soc., Atlanta, Georgia, 3-6 January 1978, pp. 195 - 208.

Collins, A.L. and T.A. Anderson. 1999. The role of food availability in regulating reproductive development in female golden perch. Journal of Fish Biology, 55 (1) : 94 - 102; 25 ref.

Colowick, S.P. and Kaplan, N.O. 1955. Lipase titremethods using water soluble substrates. In : Method in Enzymology, I : 630 - 631, ed. Colowick, S.P. and Kaplan, N.O., Academic press, New York.

Congdon, J.D. and Tinkle, D.W. 1982. Energy expenditure in free-ranging sagebrush lizards (*Sceloporus graciosus*). Canadian Journal of Zoology, 60 : 1412-1416.

Corner, E.D.S., Conwey, C.B. and Marshall, S.M. 1965. On the nutrition and metabolism of zooplankton. III. Nitrogen excretion by *Calanus*. J. Mar. biol. Ass. U.K., 45 : 429 - 442.

Costlow, J.D.1968. Metamorphosis incrustaceans. In : Metamorphosis : A problem in developmental Biology, W. Etkin and L.I. Gilbert (eds), Appleton - century - crofts, New York , 3 - 41.

Costlow, J.D., Jr. 1966. The effect of eyestalk extirpation on larval development of the mud crab, *Rhithropanepeus harrisid* (Gould). Gen. Comp. Endocrin., 7 : 255 - 274.

Cressa, C. 1986. Estimaciones de peso seco en function de la longitud cefalica y clases de tamano en *Campsurus* Sp. (Ephemeroptera, polymitarcidae). Acta Cient. Venez. 37 : 170-173.

Cui, Y. and R.J. Wootton. 1988. Bioenergetics of growth of a cyprinid, *Phoxinus phoxinus* : The effect of ration, temperature and body size and food consumption, faecal production and nitrogenous excretion. Biol., 33 : 431 - 443.

Cuzin-Roudy, J., E. Albessard, Virtue, P. and Mayzaud, P. 1999. The scheduling of spawning with moult cycle in northern krill (Crustacea: Euphausiacea): a strategy for allocating lipids to reproduction. J. Crustacean Biol., 36: 163-170.

Dall, W. 1977. Review of the physiology of growth and moulting in rock lobsters. Circ. CSIRO Div. Fish. Oceanogr., (Aust.), 7 : 75 - 81.

Dall, W. 1975. Blood carbohydrates in Western rock lobster, *Panulirus longipes* (Milne - Edwards). J. Exp. Mar. Biol. Ecol., 18, 227 - 238.

Dall, W., Hill, P.G., Rothilsburg, B.J. and Staples, D.J. 1990. The Biology of Penaeidae. Adv. Mar. Biol., 27 (1-XIII) : 1- 489.

Davenport, J., Kjorsvik, E. and Haug, T. 1990. Appetite, gut transit, oxygen uptake and nitrogen excretion in captive Atlantic halibut, *Hippoglossus hippoglossus* L., and lemon sole, *Microstomus Kitt* (Walbaurn). Aquaculture, 90: 267 - 277.

Davies, R.W., Wrona, F.J. and Kalarani, V. 1992. Assessment of activity - Specific metabolism of aquatic organisms : an imporved system. Canadian journal of fisheries and aquatic sciences, 49 : 1142 - 1148.

Davis, G.E. and Warren, C.E. 1968. Estimation of food consumption rates. IBP (International Biology Programme) Handbook, 3 : 204 - 225.

Davis, J.P. and Wilson, J.G. 1985. The energy budget and population structure of *Nucula turgida* in Dublin Bay. Journal of Animal Ecology, 54 : 557 - 571.

Dean, J.M. and Vernberg, F.J. 1965. Effect of temperature acclimation on some aspects of carbohydrate metabolism in decapod crustacea. Biol. Bull., 129 (1) : 87 - 94.

Deshmukh, R.D. 1968. Some aspects of biology of the marine crab, Scylla serrata (Forskal). Ph.D., Thesis. University of Bombay, Bombay, India.

Diana, J.S. 1983. An energy budget for northern pike, Esox lucius. Can. J. Zool., 61 : 1968.

Dircksen, H. 1992. Fine structure of the neurohemal sinusgland of the shore crab carcinus maenas L. and immuno - electron Microscopic identification of neurosecretory endings according to their neuropeptide contents. Cell Tissue Res., 269 : 249 - 266.

Drach, P. and Tchernigovtzeff, C. 1967. Surla method de determination des stades d intermue et son application generate aux crustace's. Vie Milieu, Ser. A : Biol Mar., 18 : 595 - 610.

Drach, P. 1939. Mve et cycle d□intermve chez les crustaces Decapodes. Annls. inst. oceanogr., 19 : 103 - 391.

Drach, P. 1944. Etude preliminare surle cycle d□intermue et son conditionnement hormonal chez *Leander serratus* (pennant). Biol. Bull., 78 : 40 - 62.

Du Preez, H.H., H.Y. Chen and C.S. Hsieh. 1992. "Apparent specific dynamic action of food in the grass shrimp, *Penaeus monodon* Faricius", Comp. Biochem. Physiol., 103A : 173 - 178.

Dunlop, D.S., W.V. Elden and A. Lajtha. 1978. Protein degradation rates in regions of the CNS *In vivo* during development. Biochem. J., 170, 637.

Duronslet, M.J., Yudin, A.I., Wheeler, R.S. and Clark, W.H. 1975. Light and fine structural studies of natural and artificially induced egg growth of penaeid shrimp. Proc. Annu. Workshop, World Maricult. Soc., 6 : 105 - 122.

Eastman-Reks, S. and Fingerman, M. 1984. Effects of neuroendocrine tissue and cyclic AMP on ovarian growth in vivo and in vitro in fiddler crab, *Uca pugilator*. Comp. Biochem. Physiol., 79A : 679 - 684.

Edsall, T.A., E.H. Brown, Jr. T.G. Yocum and Jr. R.S.C. Wolcott, Jr. 1974. Utilization of alewives by coho salmon in lake Michigan. U.S. Fish and Wildlife Service, Great Lakes Fishery Laboratory, Administrative Report, Ann Arbor., Michigan.

Eggers, D.M. 1977. Factors in interpreting data obtained by diel sampling of fish stomachs. Journal of the Fisheries Research Board of Canada, 34 : 1018 - 1019.

Elliott, J.M. 1976. Energy losses in the waste products of brown trout (*Salmo trutta* L.) J. Anim. Ecol., 45 : 561 - 580.

Elliott, J.M. and Davison, W. 1975. Energy equivalents of oxygen consumption in animal energetics. Occologia, 19 : 195-201.

Elliott, J.M. 1979. Energetics of fresh water teleosts. Symp.Zool.Soc., Lond., 44 : 29 - 61.

Elliott, J.M. and L. Persson, 1978. The estimation of daily rates of food consumption for fish. J. Anim. Ecol., 47 : 977.

Emmerson, W.D. 1980. Induced maturation of prawn *Penaeus indicus*. Mar. Ecol. Prog. Ser., 2 : 121 - 131.

Ennis, G.P. 1972. Growth per moult of tagged. Labsters *Homarus Americanus* in Bonavista Bay, New foundland. J. Fish. Res. Bd., Canada, 29 : 143 - 148.

Erribabu, D. and Hanumantha Rao, K. 1983. Some pre-ecdysial and post-ecdysial changes of the hepatopancreas in *Menippe rumphii* (Fabricius). Indian J. Comp. Anim. Physiol., I - 2 : 53 - 60.

Farrell, J. and Campana, S.E. 1996 : Regulation of calcium and strontium deposition on the otoliths of juvenile tilapia *Oreochromis niloticus*. Comparative Biochemistry and Physiology, 115 A : 103-109.

Fingerman, M.T., Dominiczak, M., Miyawaki, C., Oguro, C. and Yamamoto, Y. 1967. Neuroendocrine control of the hepatopancreas in the crayfish, *Procambarus clarkii*. Physiol. Zool., 40 : 23-30.

Fingerman, M. 1997. Roles of neurotransmitters in regulating reproductive hormones release and gonadal maturation in decapod crustaceans. Intertebr. Reprod. Dev., 31 : 47 - 54.

Fingerman, M., Nagabhushanam, R., Sarojini, R. and Reddy, P.S., 1994. Biogenic amines in crystaceans : identification, localization, and roles. J. Crustacean Biol., 14 (3) ; 413 - 437.

Fischer, Z. 1973. The elements of energy balance in grass carp (*Ctenopharyngodon idella*). Part IV. Consumption rate of grass carp fed on different types of food. Polsk. Arch. Hydrobiol., 20: 309 - 318.

Flint, R.W. 1972. Effects of eyestalk removal and ecdysterone infusion on moulting in *Homarus americanus*. J. fish. Res. Bd. Canada, 29 : 1229 - 1233.

Florkin, M. 1966. Aspects of moleculaires de l'adaptation et de la phylogenie. Masson, Paris.

Florkin, M. and Schoffeniels, M. 1969. Molecular Approaches to Ecology. Academic Press, New York.

Folch, J., Less, M. andSloane - Stanley, G.H. 1957. A simple method for the isolation and purification of total lipids from animal tissues. J. Biol. Chem., 226 : 497 - 509.

Fox, J.M. and Treece, G.D. 2000. Eyestalk ablation, pp. 329-331. *In*: R.R. Stickney (ed.), Encyclopedia of Aquaculture. John Wiley and Sons, Inc. ISBN 0 - 471 -29101 - 3.

Freeman, J.A. and Costlow, J.D. 1980. The molt cycle and its hormonal control in *Rhithropanoeus harristii* larvae. Dev. Biol., 74 ; 479-485.

Freeman, J.A. and Bartell, C.K. 1975. Characterization of the moult cycle and its hormonal control in *Palaemonetes pugio* (Decapoda, Caridea). Gen. Comp. Endocr., 25 : 517 - 528.

Froese, R. and Pauly, D. 1998. Fish Base 1998 : Concepts design and data sources. Manila, ICLARM. 293.

From, J. and Rasmussen, G. 1984. A growth model, gastric evacuation, and body composition in rainbow trout, *Salmo gairdneri* Richardson, 1836. Dana, 3 : 61 - 139.

Fry, F.E.J. 1971. The effects of environmental factors on the physiology of fish. In Fish physiology (Edited by W.S. Hoar and D.J. Randall). 6 : 1 - 98, Academic Press, New York.

Gabe, M. 1953. Sur l' êxistence, chez Quelques crustacês Malacostracês, d'un organe comparable a' la glande de la mue des insects. C.R. Acad. Sci. Paris, 237 : 1111 - 1113.

Gasca-Leyva, J.F.E., Martanez-Palacios, C.A. and Ross,L.G. 1991. The respiratory requirements of *Macrobrachium acanthurus* (Weigman) at different temperatures and salinities. Aquaculture, 93 : 191 - 197.

Gendron, L., Fradette, P. and Godbout, G. 2001. The importance of rock crab (Cancer irroratus) for growth, condition and ovary development of adult American lobster (Homarus americanus). J. Expt. Mar. Biol. Ecol. 262 : 221-241.

George, M.J. 1972. On the food and feeding habits of *Metapenaeus monoceros* (Fabriccius). Fishing News (Books) Ltd., London, 178.

George, M.J. 1976. The food and feeding of the shrimp *Metapenaeus monoceros* (Fabricius). Caught from the backwaters. Indian J. Fish., 21(2) : 495 - 500.

Gerard, J.F., and Gilles, R. 1972. Free amino acid pool in *Callinecles scpidus*, (Rathbum), tissues and its role in the osmotic intracellular regulation. J. Exp., Mar. Biol. Ecol., 10: 125 - 136.

Gerking, S.D. 1955. Endogenous nitrogen excretion of bluegill sunfish. Physiol. Zool., 28 : 283 - 289.

Germano, B.P. 1994. Effects of unilateral eyestalk ablation on growth in juvenile blue crabs *Portunus pelagicus* (L.) (Crustacea : Decapoda : Portunidae). Asian. Fish. Sci., 7 (1) : 19 - 28.

Gilbert, L.I. and O Connor, J.D. 1970. Lipid metabolism and transport in arthropoda, In : Chemical Zoology edited by Florkin, M. and Scheer, B.T., Academic Press, New York, 5 : 229 - 254.

Gilbert, L.I. 1969. Proc. 3rd Intern. Congr. Endocrinol., Mexico City 1968. Intern Congr. Ser. No. 157. pp. 340 - 346. Excerpta Med. Found, Amsterdam.

Gilbert, L.I. 1967. Advan. Insect. Physiol., 4 : 69.

Gilles, R. 1969. Effect of various salts on the activity of enzymes implicated in amino acid metabolism. Archs in Physiol. Biochem., 77 : 441 - 464.

Glencross, B.D., Smith, D.M., Tonks, M.L., Tabrett, S.M. and Williams. K.C. 1999. A reference diet for nutritional studies of the prawn, Penaeus monodon. Aqua., Nutr., 5 : 33 - 40.

Goldberg, A.L. 1974. Intracellulor protein degradation in mammalian and bacterial cells. Ann. Rev. Biochem. 43, 835.

Gomez, R. 1965. Acceleration of development of gonads by implantation of brain in the crab *Paratelphusa hydrodromous*. Naturwissen Schaften, 9 : 216 - 221.

Gonzalez, M.L., Perez, M.C., Lopez, D.A. and Buitano, M.S. 1990. Effect of temperature in the energy availability for growth of *Concolepas concolepas* (Brugiere). Rev. Biol. Mar., 25 : 71 - 81.

Gordon, M.S. 1972. "Animal physiology : Principles and Adaptations", 2nd Ed. Macmillan, New York.

Gould, S.1966. Allometry and size in Ontogeny and Phylogeny. Biological Research, 41 : 587 - 640.

Gould, S.J. 1992. This was a man. Foreword to abridged Canto edition On Growth ansd Form (ed. J.T. Bonner). PP. IX-XIII. Cambridge : Cambridge University Press.

Goulden, C.E. and place, A.R. 1990. Fatty acid synthesis and accumulation rates in daphniids, J. Exp. Zool. 256 : 168 - 178.

Grainde, B. and Seglen, P.O. 1981. Effect of amino acid analogues on protein degradation in rat hepatocytes. Biochemica. Acta. Biophysica. Acta., 676 : 43 - 50.

Griffiths, D. 1991. Food availability and the use and storage of fat by ant-lion larvae. Oikos, 60 : 162 - 172.

Grime, J.P. 1989. The stress debate : Symptom of impending synthesis? Biological journal of the linnean society, 37 : 3 - 17.

Guisande, C and Serrano, L. 1989. Analysis of protein, carbohydrate and lipid in rotifers. Hydrobiologia, 186 : 339 - 346.

Guy Selman, J.B. 1953. An analysis of the moulting process in the fiddler crab, *Uca pugilator*. Biol. Bull., 104 : 115 - 137.

Guyton, C. 1981. Lipid metabolism. In ; Text Book of Medical Physiology (6th edn.), W.B. Saunders Company, Philadelphia, London, Toronto, Igakushon Ltd., Tokyo, 849-859.

Haiatt, R.W. 1948. The biology of the lined shore crab, *Pachygrapsus crassipes* Randall. Pac. Sci., 2 : 135 - 213.

Hainsworth, F.R., Tardiff, M.F and Wolf, L.L. 1981. Proportional Control for daily energy regulation in humming birds, Physiological Zoology 54 : 452-462.

Handy, R.D., Sims, D.W., Giles, A., Campbell, H.A. and Musonda, M.M. (1999). Metabolic trade-off between locomotion and detoxification for maintenance of blood chemistry and growth parameters by Rainbow Trout (*Oncorhynchus mykiss*) during chronic dietary exposure to copper. Aquatic Toxicology, 47: 23-41

Hanel, R., Karjalainen, J and Wieser, W. 1996 : Growth of swimming muscles and its metabolic cost in larvae of whitefish at different temperatures. Journal of Fish Biology, 48 : 937 - 951.

Hanstorm, B. 1931. Neue untersuchangen uber sinnes organe and Nervensystem der crustacean. Z. Morphol. Tiere., 23 : 80 - 236.

Harper, H.A., Rodwell, V.M. and Mayer, P.A. 1979. In : Review of Physiological Chemistry, 17th edition, Lange Medical Publications, Muruzer Company Limited, California.

Harper, H.A. 1983. In : Harper's Review of Biochemistry (Eds. D.W. Martin, P.A. Mayes and V.W. Rodwell 19th Edn.), Lange Medical Publications, Maruzen, Asia, Singapore.

Harper, H.A. 1971. Review of physiological chemistry, 13th edition. Lange Medical publications. Los Altos, California.

Hartenstain, R. 1970. In comparative biochemistry of nitrogen metabolism. I. The invertibrates. Ed., Cambell, J.W. Academic press, New York, 299.

Hartnoll, R.G. 1973. Factors affecting the distribution and behaviour of the crab *Dotilla fenestrata* on East African shores. Estuarine Coastal Mar. Sci., 1 : 137 - 152.

Hartnoll, R.G. 1974. Variations in growth pattern between some secondary sexual characters in crabs (Decapoda : Brachyura). Crustaceana, 27 : 131 - 136.

Henderson, P.A., H.A. Holmes, and R.N. Bamber. 1988. Size-selective overwintering mortality in the sand smelt, *Atherina boyeri* Risso, and its role in population regulation. Journal of Fish Biology, 33:221-233.

Heroux, D. and Magnan, P. 1996. In situ determination of food daily ration in fish ; review and field evaluation. Environmental Biology of Fishes, 46 : 61 - 74.

Herreid, C.F. 1980. Hypoxia in invertebrates. Comp. Biochem. Physiol., 67A : 311 - 320.

Hershko, A. and A.Ciechanover. 1982. Mechanisms of intercellular breakdown. Ann. Rev. Biochem, 51, 335.

Hewett, S.W. and Johnson, B.L. 1992 : A generalized bioenergetics model of fish growth for microcomputers - university of Wisconsin Sea Grant Technical Reports No. WIS - SG - 91 - 250. Second edition.

Highnam, K.C. and Hill, L. 1977. Endocrine mechanisms in crustacea. In : The comparative endocrinology of the invertebrates. ELBS edn. Edward Arnold (Publishers) Ltd., London, 209 - 257.

Hill, C., Quigley, M.A., Cavaletto, J.F and Gordon, W. 1992. Seasonal changes in lipid content and composition in the benthic amphipods *Monoporeia affinis and pontoporeia femorata*. Limnol. Oceanogr., 37 (6), 1280-1287.

Hinsch, G.W. 1977. Fine structural changes in the mandibular gland of the male spider crab, *Libinia emarginata* (L.) following eyestalk ablation. Journal of morphology, 154 : 307 - 315.

Hinsch, G.W. and Bennet, 1979. Vitellogenesis stimulated by thoracic ganglion implants into destalked immature spider crabs, *Libinia emarginata*. Tissue and Cell, 11 : 345 - 351.

Hoar, W.S. 1984. General and Comparative Physiology, Prentice-Hall of India, Pvt. Ltd. New Delhi.

Hofer, R., G. Krewedl and F. Koch. 1985. An energy budget for an omnivorous cyprinid, *Rutilus rutilus* (L.) Hydrobiologia, 122 : 53 - 59.

Hohnke, L. and Scheer, B.T. 1970. Carbohydrate metabolism in crustaccans. In : Chemical Zoology, Vol.V, Arthropod Part A, (Eds). M. Florkin and B.T. Scheer, Academic Press, New York and London, 147 - 165.

Holthuis, L.B. 1980. FAO species catalogue. Vol.1. Shrimps and prawns of the world. An annotated catalogue of species of interest of fisheries. FAO fisheries Synopsis, No. 125 (1) : 1 - 26.

Homola, E. 1997. Distribution and regulation of esterases that hydrolyze methylfarnesoate in *Homarus americanus* and other crustaceans. General and Comparative Endocrinology, 106 : 62 - 72.

Hornung, D.E. and Stevenson, J.R. 1971. Changes in the rate of Chitin synthesis during the crayfish molting cycle. Comp. Biochem. Physiol., 40 B : 34-346.

Hu, A.S.L. 1958. Glucose metabolism in *Carcinus maenas*. Comp. Biochem. Biophys., 75 : 387 - 395.

Huberman, A. 2000. Shrimp endocrinology. A review. Aquaculture, 191 : 191 - 208.

Huggins, A.K. and Munday, K.A. 1968. Crustacean metabolism. In : Advances in Comparative Physiology and Biochemistry, Vol. 3 (ed) O. Lowenstein, Academic press, New York, London, 271 - 377.

Huggins, A.K. 1966. Intermediatry metabolism in *Carcinus maenas*. Comp. Biochem. Physiology, 18 : 283 - 290.

Hughes, G.M. 1984. Scaling of respirtion areas in relation to oxygen consumption of vertebrates. Experientia, 40 : 519 - 524.

Hunn, J.B. 1982. Urine flow rate of fresh water of salmonids: A mini - review. Prog. Fish. Cult., In press.

Huxley, J.S. 1932. Problems of Relative growth. London : Methuen and Co. Ltd.

Idyll, C.P. 1971. Induced maturation of ovaries and ova in pink shrimp. Comm. Fish. Rev., 33 : 1 - 20.

Ikeda, T. 1985. Metabolic rates of epipelagic marine zooplankton as a function of body mass and temperature. Mar. Biol., 85 : 1 - 11.

Itazawa, Y. and Oikawa, S. 1986. A quantitative interpretation of the metabolism -size relationship in animals. Experientia, 42 : 152 - 153.

Iwata, K. 1970. Relationship between food and growth in young crucian carps, *Carassius auratus* curvieri, as determined by the nitrogen balance. Jpn. J. Limnol., 31 : 129 - 151.

Jaeger, R.G. and Barnard, D.E. 1981. Foraging tactics of a terrestrial salamander : choice of diet in structurally simple environments. American naturalist, 117 : 639-664.

Jauncey, K. 1982. Carp *(Cyprinus Carpio L.)* nutrition- A review cited in Recent Advances in Aquaculture, edited by J.F.Muir and R.J.Robert, west view press, Inc. 5500, Colarado, pp : 222.

Jayasundaramma, B. and Ramamurthi, R. 1988. Aspects of neuroendocrine control of metabolism in the freshwater rice field crab, *Oziotelphusa senex senex.* IV. Aspects of protein metabolism during the induced molt cycle. Trends In life Science (India), 3 (2) : 39-45.

Jhingran, V.G. 1991. Fish and Fisheries of India. Hindustan Publishing Corporation (India). New Delhi, India.

Jobling, M. 1985. Growth. In : Tytler, P. and Calow, P. (eds). Fish energetics : New perspectives, pp. 213 - 230. Croom Helm, London.

Jobling, M. 1994. Fish bioenergetics, Chapman and Hall, London, 309.

John, M.C. 1957. Bionomics and life history of *Macrobrachium rosenbergii* (de Man), Bull. Central Res. Inst. University of Kerala. Ser., 5 (1) : 93 - 102.

Jones, C. M. 1992 : Development and application of the otolith increment technique. In : Stevenson, D.K., Campana, S.E. (eds) : Otolith micro structure examination and analysis. Canadian Special Publication of Fisheries and Aquatic Sciences, 117 : 1 - 11.

Josse, J. and W.P.M. Geraerts. 1983. Endocrinology. pages 317 - 406 in A.S.M. Saleuddin and K.M. Wilbur, editors. The mollusca: physiology, Vol.4. Academic press, New York, USA.

Jugan, P. and Van Herp, F. 1989. Introductory study of an oocyte membrane protein that specially binds vitellogenim in the crayfish, *Orconectus limosus*. Invert. Reprod. Devel., 16 ; 149-154.

Juinio - Menez, M.A. and Ruinata, J. 1996. Survival, growth and food conversion efficiency of panulirus ornatus following eyestalk ablation. Aquaculture, 146 : 225 - 235.

Kamemoto, F.I. 1976. Neuroendocrinology of osomoregulation in decapod crustacea. Am. Zool., 16 : 141-150.

Kamiguchi, Y. 1971. Studies on the moulting in the freshwater prawn *Palaemon paucidens* 1. Some endogenous and exogenous factors influencing the intermoult cycle. J. Facult. Sci., Hokkaido Univ., 18 : 15 - 23.

Karplus, I. and Hulata, G. 1995. Social control of growth in *Macrobrachium rosenbergii*. V. the effect of unilateral eyestalk ablation on jumpers and laggards, Aquaculture, 138 : 181 - 190.

Kaufmann, K.W. 1981. Fitting and using growth curves. Oecologia., Berlin, 49 : 292 - 299.

Keckeis, H. and Schiermer, F. 1990. Consumption, growth and respiration of bleak, *Alburnus* alburnus (L.), and roach, Rutilus rutilus (L.), during early ontogency. Journal of Fish Biology, 36 : 841 - 851.

Keller, R. and Sedlmeier, D. 1988. Metabolic hormone in crustaceans : the hyperglycemic neuropeptides. In : Laufer H, Downer RGH, eds. Endocrinology of selected invertebrate types. 12 : 315-357.

Keller, R. 1966. Uber eine hormonale regulation der glycogen synthese beim fluss Kerbs. *Orconectes linosus*. Verh. d. Zool. Ges., 18 : 272 - 279.

Keller, R. 1992. Curstacean neuropeptides : Structure, functions and comparative aspects. Experientia. 48 : 439 - 448.

Kelso, J.R.M. 1972. Conversion, maintenance and assimilation for walleye, *Stizostedion vitreum vitreum*, as affected by size, diet and temperature. J. Fish. Res. Board. Can., 29 : 1181 - 1192.

Kemp, M.B.A. and Mayers, D.K. 1954. A Calorimetric micromethod for the determination of glucose. Biochem. J., 56 : 639 - 645.

Kewalraman, H.G., Shankolli, K.N. and Shenoy, S.S. 1991. On the larval life history of *M. malcolmsonii* under captivity. J. Indian. Fish Asoc. 1 (1) : 1 - 25.

Khalaila, I., Well, S. and Sagi, A. 1999. Endocrine balance between male and female components of the reproductive system in intersex cherax quadricarinatus (Decapoda : Parastacidae). J. Expt. Zool., 283 : 286 - 294.

Kiron, V. Diwan, A.D. 1985. Influence of eyestalk ablation on ammonia excretion in the prawn *Penaeus indicus* (H. Milne Edwards). Indian J. Mar. Sci., 14 (4) : 220 - 221.

Kishori, B., Premasheela, B., Ramamurthy, R. and Sreenivasula Reddy, P. 2001. Evidence for a Hyperglycemic effect of Methionine enkephalin in the prawns *Penaeus indicus* and *Metapenaeus monocerus*. General and comparative endocrinology, 123 : 90 - 99.

Kitchell, J.F., D.J. Stewart, and D. Weininger. 1977. Applications of a bioenergetics model to yellow perch (*Perca flavescens*) and walleye (*Stizostedion vitreum vitreum*). J. Fish. Res. Board. Can., 34: 1922 - 1935.

Kleiber, M. 1961. The fire of life. An introduction to animal energetics. John Wiley, New york.

Kleine, R. 1967. Z. Vergleich. Physiol., 55 : 333.

Klekowsky, R.Z. and Duncan, A. 1975. Physiological approach to ecological emergetics. In w. Grodzinski, R.Z. Klekowsky and A. Duncan (eds). Methods for ecological bioenergetics. Black well Sci. Pub., Oxford. 15 - 59.

Knowels, F.G.W. and Carlisle, D.W. 1956. Endocrine control in the crustacea. Biol. Rev., 31 : 396 - 473.

Koshio, S., Teshima, S. and Kanazawa, A., 1992a. Effect of unilateral eyestalk ablation and feeding frequencies on growth, survival, and body compositions of juvenile freshwater prawn *Macrobrachium rosenbergii*. Bull. Jpn. Soc. Sci. Fish., 58 : 1419 - 1425.

Koshio, S., Castell, J.D. and O' Dor, R.K. 1992b. The effect of different dietary energy levels in crabs -protein-based diets on digestibility, oxygen consumption and ammonia excretion of bilaterally eyestalk ablated and intact juvenile lobsters, *Homarus americanus*. Aquaculture, 108 (3-4) : 285 - 297.

Kremer, P. 1977. Respiration and excretion by the ctenophore, Mnemiopsis leidyi. Mar. Biol., 44 : 43 - 55.

Kuo, C.M. Hsu, C.R. and Lin, C.Y. 1995. Hyperglycemic effects of dopamine in tiger shrimp, *Penaeus monodon*. Aquaculture, 135 : 161 - 172.

Kurup, N.G. 1964. The intermoult cycle of an anomuran *Petrolisthes cinctipes* Randall. Biol. Bull., 127 : 97 - 107.

Lafon, M. 1948. Nouvelles Researches bio chimiques et physiologiques sur le squelette tegumentaire des crustaces. Bull. Inst. Oceanog., 45 : 1 - 28.

Langer, R.K. and Somalingam, 1993. Experimental culture of freshwater process in ponds at powarkheda, Hoshangabad (M.P.). III Indian fisheries forum putnagar, Oct. 11 - 14, 1993, Abst. No.12.

Larson, R.J. 1991. Seasonal cycles of reserves in relation to reproduction in Sebastes. Env.Biol. Fish. 30, 57 - 70.

Laufer, H. Brost, D.W. Baker, F.C., Carrasco, C., Sinkus, M., Recuter, C.C., Tasi, L.W. and Schooley, L.A. 1987. Identification of juvenile hormone- like compound in a crustacean. Science, 235 : 202 - 205.

Laufer, H., Landau, M., Borst, D.W. and Homola, E. 1986. The synthesis and regulation of methylfarnesoate, a new juvenile homrone for crustacean reproduction. In : M. porchet, J.C. Andries and A. Dhainaut (Eds) Advances in invertebrate Reproduction 4. Elesevier Science Publishers, Amsterdam, pp. 135 - 143.

Laverdure, A.M. and Soyez, D. 1988. Vitellogemin receptor from lobster oocyte membrane : Solubilization and characterization by solid phase binding assay, Int. J. Invert. Reprod. Devel., 13 ; 251-266.

Lawrence, J.M. 1976. Patterns of lipid storage in post □ metamorphic marine invertebrates □ Am.Zool. 16:747-762.

Le Roux, A., 1984. Quelques effects de l ablation des pédoncules oculaires sur les larges et les premiers stades juvéniles de de Palaemonetes varians (Leach) (Decapoda, Palaemonidae). Bull. Soc. Zool. Fr., 109 : 43-60.

Lee, K.K., Chen, Y.L. and Liu, P.C. 1999. Hemostasis of tiger prawn. *Penaeus monodon* affected by *Vibrio harveyi,* Extracxellular products and toxic cysteine protease. Blood cells, Moleules, and Diseases, 15(13) : 180-192.

Lee, Y.L. and Lardy. H.A. 1965. Influence of thyroid hormones on L-glycerophosphate dehydrogenases and other dehydrogenases in various organs of the rat. J.Biol. Chem., 240 : 1427 - 1432.

Lee, R.F. and Puppione, D.L. 1988. Lipoproteins I and II from the hemolymph of the blue crab, *Callinecter Sapidus*, lipoprotein II associated with vitellogenins. J. Expt. Zool., 248 : 278-289.

Lehninger, A.L. 1984. Principles of Biochemistry, The Johns Hopkins University School of Medicine, 1st Indian edn., C.B.S. Publishing Co., New Delhi.

Lei, C.H., L.Y. Hsieh and C.K. Chen. 1989. Effects of salinity on the oxygen consumption and ammonia excretion of young juveniles of the grass shrimp, *Penaeus monodon* Fabricius. Bull. Inst. Zool., Academia Sinica 28 : 245 - 256.

Liao, I.C. and H.J. Huang. 1975. Studies on the respiration of economic prawn in Taiwan. 1. Oxygen consumption and lethal dissolved oxygen of egg up to young prawn of *Penaeus monodon* Fabricius. J. Fish. Soc., (Taiwan) 4 : 33 - 50.

Liao, I.C. and T. Murai. 1986. Effects of dissolved oxygen, temperature and salinity on the oxygen consumption of the grass shrimp, *Penaeus monodon*. In the first Asian Fisheries Forum (Edited by J.L. Maclean, L.B. Dizon and L.V. Hosilos). pp. 641-646. Asian fisheries society, Manila, Philippines.

Lied, E., Lund,B. and Vonder Decken,H. 1982. Protein synthesis in vitro by expaxial muscle polyribosomes from cod, *Gadus morhua.* Comparative Biochemistry and physiology. 72 (B) : 187 - 193.

Limburg, K.E., Pace,M.L. and Arend, K.K. 1998. Growth and recruitment of larval *Morone* in relation to food availability and temperature in the Hudson River. Fish. Bull. 97: 80-91.

Lin, C.Y., Chen, S.H., Kou, G.H. and Kuo, C.M. 1998. Identification and characterization of a hyperglycemic hormone from freshwater giant prawn, *Macrobrachium rosenbergii*. Comparative Biochemistry and Physiology, 121 (A) ; 315-321.

Ling, S.W. 1962. Studies on the culture of the giant freshwater prawn *Macrobrachium rosenbergii* (de Man). IPFC curr. affairs Bull., 35 : 1 - 11.

Ling S.W. 1969. The biology and development of *M.rosenbergii*. FAO. Fish Rep., 57 (3) : 589 - 606.

Lio, I.C., Ting, Y.Y. and Kalsutani, K.K. 1969. Summary of Preliminary report on artificial propagation of *Metapenaeus monoceros* Fab. Joint Commission on Rural Construction. Fish. Ser., 8 : 72 - 76.

Little, G. 1969. The larval development of shrimp, *Palaemon macrodactylus* Rathbun, reared in the laboratory, and the effect of eyestalk extirpation on development. Crustaceana, 17 : 69 - 87.

Liu, L. Laufer, H. 1996. Isolation and characterization of sinus gland neuropeptides with both mandibular organ inhibiting and hyperglycemic effects from the spider crab *Libinia emarginata*. Arch. Ins. Biochem. Physiol., 32 : 375 - 385.

Liu, R.Y. 1983. Shrimp mariculture studies in China. In proceedings of the first international conference on warm water Aquaculture - Crustacean (Edited by G.L. Rogers, R. Day and A. Lim), pp. 82 - 87. Brigham Young University, Hawaii Campus. Laie.

Logan, D.T. and Epifanio, C.E. 1978. A laboratory energy balance for the larvae and juveniles of the American lobster, *Homarus americanus*. Mar. Biol., 47 : 381- 389.

Lowry, O.H., Rosenbrough, J.J., Farr, A.L. and Ran dall, R.J. 1951. Protein measurement with the foline phenol reagent. J. Biol. Chem., 193 : 265 - 275.

Lumare, F. 1979. Reproduction of *Penaeus kerathurus* using eyestalk ablation. Aquaculture, 18 : 203 - 214.

MacLean, M. H., Ang, K. J, Brown, J.H, Jauncey, K. and Fry, J. C. 1994. Aquatic and benthic bacteria responses to feed and fertiliser application in trials with the freshwater prawn, *Macrobrachium rosenbergii* (de Man), *Aquaculture* 120: 81-93.

Madhyastha, M.N., Rangneker, P.V. 1976. Metabolic effects of eyestalk removal in the crab, *Varuna litterata*. (Fabricus). Hydrobiolgia, 48 : 25.

Makinouchi, S. and Primavera, J.H. 1987. Maturation and spawning of *Penaeus indicus* using different ablation methods. Aquaculture, 62 : 73 - 81.

Man, K.H. 1978. Estimating the food consumption of fish in nature. In: Gerking, S.D.(ed.). Ecology of freshwater fish productin, pp. 250 - 73. Oxford, Blackwell.

Mangum, C.P., Van,W. and Winkle. 1973. Responses of aquatic invertebrates to declining oxygen conditions. Am. Zool., 13 : 529 - 541.

Mantel, L.H. and Farmer, L.L. 1983. Osmotic and ionic regulation. In : The Biology of crustacea, Vol. V. Internal anatoomy and physiological regulation. Eds. by L.H. mantel. Academic Press. New York. London, 53-161.

Marangos, C., E. Alliot, C.H. Brogren and H.J. Ceccaldi. 1990. Nycthermal variations of ammonia excretion in *Penaeus japonicus* (Crustacea, Decapoda, Penaeidae). Aquaculture. 84: 383-391.

Marian, M.P., Pandian, T.J., Mathavan, S., Murugadass, S. and Premkumar, D.R.D. 1986. Suitable diet and optinum feeding frequency in the eyestalk ablated prawn, *Macrobrachium lamarrei*, p. 589-592. In J.L Maclean, L.B. Dizon and L.V. Hosillos (eds.) The First Asian Fisheries Forum. Asian Fisheries Society, Manila, Philippines.

Martin, D.W., P.A. Mayes and V.W. Rodwell. 1981. Harper□s Review of Biochemistry., 8th Ed. Lange Medical Publications, California.

Mattson, M.P. and Spaziani, E. 1986. Regulation of crab Y-organ steriodogenesis in vitro : evidence that ecdysteroid production increases through activation of CAMP-phosphodiesterase by calcium-calmodulin. Cell. Endocrinol., 48 : 135 - 151.

Mattson, M.P., and Spaziani, E., 1985. 5-Hydroxytryptamine mediates release of molt-inhibiting hormone activity from isolated crab eyestalk ganglia, Biol. Bull. 169 : 246 - 255.

Mattson, M.P. 1986. New insights into neuroendocrine regulation of the crustacean moult cycle. Zool. Sci., 3 : 733 - 744.

Mauchline, J. 1976. The Hiatt growth diagram for Crustacea. Mar. Biol., 35: 79 - 84.

Mauviot, J.C. and Castell, J.D. 1976. Moult and growth enhancing effects of bilateral eyestalk ablation on juvenile and adult

American lobster (*Homarus americanus*). J. Fish. Res. Bd. Canada. 33 : 1922 - 1929.

Mayzaud, P. and Conover, R.J. 1988. O : N atomic ratio as a tool to describe zooplanktonmetabolism. Mar. Ecol. Prog. Ser., 45 : 289 - 302.

McWhinne, M.A. and Mohrherr, C.J. 1970. Influence of eyestalk factors, intermolt cycle and sesen upon ^{14}C- Leucine incorporation into protein in the crayfish, *Oroconectes virilis*. Comp. Biochem. Physiol., 34 : 415-437.

McWhinne, M.A. and Saller, P.N. 1960. Analysis of blood sugars in the crayfish *Orconectes virilis* . Comp. Biochem. Physiol., 1 : 110 - 122.

McWhinne, M.A. 1962. Gastrolith growth and calcium shifts in the freshwater crayfish, *Orconectes virilis*, Comp. Biochem. Physiol., 7 : 1-14.

Meenakshi, V.R., and Scheer, B.T. 1961. Metabolism of glucose in the crab, *Cancer magister* and *Hemigraspus nudus*. Comp. Biochem. Physiol., 3 : 30 - 41.

Megusar, F. 1912. Experimente uber den Farbwechel der crustaceen. Arch. Entwick lungs - mech. organ, 33 : 462 - 665.

Meyer, E. 1990. Levels of major body compounds in nymphs of the stream mayfly *Epeorus sylvicola* (Pict.) (Ephemeroptera: Heptageniidae). Arch. Hydrobiol. 117: 497-510.

Millamena, O.M. and Quinito, E.T. 2000. The effect of diets on reproductive performance of eyestalk ablated and intact mud crab *Scylla serrata*. Aquaculture, 181: 81 - 90.

Mohamed, K.S. and Diwan, A.D. 1991. Neuroendocrine Regulation of ovarian maturation in the Indian white prawn *Penaeus indicus* H. Milne Edwards. Aquaculture, 98 : 381 - 393.

Moore, S. and Stein, W.H. 1954. In : Methods in Enzymology, edited by collowick, S.P. and Kalpan, N.O., Vol. II. Academic Press, New York, 501.

Mootz, C.A. and Epifanio, C.E. 1974. An energy budget for *Menippe mercenaria* larvae fed Artemia nauplii. Biol. Bull., 146 : 44 - 55.

Morand, P., Carre, C. and Biggs, D.C. 1987. Feeding metabolism of the jellyfish *Pelagia noctiluca*. J. Plankton Res., 9 : 651 - 665.

Morris, R.J. 1984. The endemic faunae of Lake Baikal: Their general biochemistry and detailed lipid composition. Proc. R. Soc. Lond. Ser. B 222: 51-78.

Muir, B.S. and A.J. Niimi. 1972. Oxygen consumption of the euryhaline fish aholehole (*Kuhlia sandvicensis*) with reference to salinity, swimming and food consumption. J. Fish .Res.Board. Can., 29 : 67 - 77.

Muriana, F.J.G., Ruiz - Gaterrez,, V., Gallardo, M.L. and Minguez Mosquera, M.I. 1993. A Study of the lipids and carotenoprotein in the prawn *Penaeus japonicus*. J. Biochem., 114 : 223-229.

Murray, R.K., D. K.Granner, P.A.Mayes and V.W.Rodwell. 2000. Harper's Biochemistry. 25th Edition, McGraw-Hill Companies, USA.

Muthu, M.S. and Laxminarayana, A. 1982. Induced maturation of penaeid prawns-a review. Proc. Symp. Coastal Aquaculture, 1 : 16 - 27.

Muthu, M.S. 1981. Development and culture of penaeid larvae a review in Progress in invertebrate Reproduction and Aquaculture (Subramoniam, T., and Sudha Varadarajan, eds.). Proceedings of the first All India Symposium on Investebrate Reproduction - Madras University, July-1980. Published for Indian Society of Invertebrate Reproduction.

Muthu, M.S. and Laxminarayana, A. 1980. Induced maturation and spawning of Indian penaeid prawns. Indian. J. Fish., 26 : 172 - 180.

Mykles, D.L. and Skinner, D.M. 1982. Crustacean muscles : atrophy and regeneration during molting, pp. 337-357. In B.M. Twarog, R.J.C. Le vine and M.M. Dewey (ed) Biology of muscles : A comparative approach. Reven Press, New York.

Nachlas, M.M, Margulius S.P. and Saligman, A.M. 1960 : J. Biol. Chem., 235 : 499.

Nagabhushanam, R. and Vasantha, N. 1971. Moulting and colour changes in the prawn, *Caridina weberi*. Hydrobiologia, 38 : 39 - 47.

Nagabhushanam, R. and Jyothi, M. 1977. Hormonal control of osmoregulation in the freshwater prawns *Caridina weberi*. J. Anni. Morphol. Physiol., 24 : 20-28.

Nagabhushanam, R. and Vasantha, N. 1971. Moulting and color changes in the prawn *Caridina weberi*. Hydrobiologia., 38 : 39 - 47.

Nakatani, I. and Otsu, T. 1979. The effects of eyestalk, leg and uropod removal on the moulting and growth of young crayfish *Procambarus clarkii*. Biol. Bull., 157 : 182 - 188.

Nakatani, I. and Otsu, T. 1981. Relations between the growth and the moult interval in the eyestalkless crayfish, *Procambarus clarkii*. Comp. Biochem. Physiol., 68A : 549 - 553.

Nalini, C. 1976. Observations on the maturity and spawning of *Metapenaeus monoceros* (Fab.) at Cochin. Indian J. Fish, 23 (1-2) : 23 - 30.

Nan, F.A., Sheen, S.S., Cheng, Y.T. and Chen, S.N. 1995. The effect of eyestalk ablation on oxygen consumption and ammonia - N excretion of juvenile shrimp *Penaeus monodon*. Zool. Stud., 34 (4) : 265 - 269.

Natelson, S. 1971. Total cholesterol procedure (Liberman Burchard reagent). In: Techniques of clinical chemistry, III edi. Thomas, CC. Publ. Spring field, Illinoids, USA, 263 - 268.

Needham, A.E. 1964. The growth process in Animals. Pitman and Sons, London.

Neiland K.A. and B.T. Scheer. 1953. The influence of fasting and of sinus gland removal on body composition of *Hemigrapsus nudus*. Physiologia. Comp. Oecol., 3: 321 - 326.

Nelson, S.G., Knight, A.W. and Li, H.W. 1977. The metabolic cost of food utilizaiton and ammonia production by juvenile *Macrobrachium resenbergii* (Crustacea : Palaemonidae). Comp. Biochem. Physiol., 57A : 67 - 72. Pergaman press. Printed in Great Britain.

Nelson, D.L. and Cox, M.M. 2000. Lehninger Principles of Biochemistry. Thrid edition. Macmillan Publishers, U.K.

New, M.B. 1976. A review of dietary studies with shrimps and prawns. Aquaculture, 9 : 101 - 144.

New. M.B. 1988. Freshwater prawns, status of global Aquaculture 1987. NACA Technical manual series. NACA, Bangkok, Thailand. 6 : 58.

Niimi, A.J. and Beamish, F.W.H. 1974. Bioenergetics and growth of largemouth bass (*Micropterus salmoides*) in relation to body weight and temperature. Can.J. Zool., 52 : 447 - 456.

Nival, P., Malara,G., Charra, R., Palazzoli, I. and Nival, S. 1974. Etude de la respiraiton et de le'xcr'etion de quelques cope'podes planctioniques (crustacea) dans la zone de remonte'e d'eau profonde des cotes Marocaines. J. Exp. Biol. Ecol., 15 : 2231 - 261.

O' Connor, J.D. and Gilbert, L.T. 1969. Alteration in lipid metabolism associated with premolt events in a land crab, and fresh water crayfish. Comp. Biochem. Physiol., 29 : 889 - 904.

O' Connor, J.D. and Gilbert, L.T. 1968. Alteration in lipid metabolism in crustaceans. Am. Zool., 8 : 529 - 539.

Obuchowicz, L. 1966. Hormonal regulation of metabolism in HP mitochondria of crayfish, *Astacus leptodactylus*. Warsaw Fedration European Biochem. Soc. Meeting., Abstr, No.1797.

Oikawa, S. and Itazawa, Y. 1984. Allometric relationship between tissue respiration and body mass in the carp. Comp. Biochem. Physiol., 77A : 415 - 418.

Oliver, J.D., G.F. Holeton and K.E. Chua. 1979. Overwinter mortality of fingerling small mouth bass in relation to size, relative energy stores and environmental temperature. Transactions of the American Fisheries Society. 108: 130-136.

Olson, R.J. and Mullen, A.J. 1986. Recent developments for making gastric evacuation and daily ration determinations. Envrionmental Biology of Fishes, 16 : 183 - 191.

Otsu, T. 1963. Bihormonal control of sexual cycle in the freshwater crab, *Potamon dehaani*. Embryologi., 8 : 1 - 20.

Pandian, T.J. and Balasundaram, C. 1982. Moulting and spawning cycles in *Macrobrachium nobilii* (Henderson and Mathai). Int. J. Invertebr. Repord., 5 : 21 - 30.

Pandian, T.J. 1989. Protein requirements of fish and prawns cultured in Asia, p. 11-22. In S.S. De Silva (ed.) Fish Nutrition Research in Asia. Proceedings of the Third Asian Fish Nutrition Network Meeting, Asian Fish. Soc. Spec. Publ., 4, 166 p. Asian Fisheries Society, Manila, Philippines.

Pandian, T.J. and Sindhu Kumari, S. 1985. Does eyestalk ablation induce hyperphagia? Curr. Sci., 54 : 1281 - 1282.

Pandian, T.J. 1987. Fish. In Animal Bioenergetics, Vol. 2. Bivalvia through Reptilia (Edited by T.J. Pandian and F.J. Vernberg), pp. 357 - 465. Academic press, San Diego.

Panikkar, N.K. 1941. Osmoregulation in some Palaemonid prawns. J. Mar. Biol. Ass. U.K., 25 : 317 - 359.

Panouse, J.B. 1943. Influence de l' ablation du pe donucle oculaire sur la croissance de l☐ ouaire chez la crevette *Leander serratus*. C.R. Acad. Sci., 217 : 553 - 555.

Panouse, J.B. 1947. La glande du Sinus et al maturation des produits genitaux chez les crevettes. Bull. Biol. Belg. (Suppl.), 33 : 160 - 163.

Parvathy, K. 1972. Endocrine regulation of carbohydrate metabolism during the molt cycle in crustaceans. I. Effect of cyestalk removal in *Ocypode plantytaris*. Mar. Biol., 14 : 58 - 62.

Passano, L.M. 1953. Neurosecretory control of moulting in crabs by the X-organ sinus gland complex. Physiol. Comparata et Oecol., 3 : 155 -189.

Passano, L.M. 1960. Moulting and its control. In : The Physiology of Crustacea, (ed. Waterman, T.H.), Academic Press, New York. 1 : 473 - 536.

Peters, R.H. 1983. The ecological implications of body size. Cambridge studies in Ecology No.2. Cambridge University Press. Cambridge.

Petrusewicz, K. and Macfadyen, A. 1970. Productivity of terrestrial animals : Principles and methods. I.B.P. Handbook No. 13, Block well scientific publications, oxford.

Pillay, K.K. and Nair, N.B. 1970. The reproductive cycle of three decapod crustaceans from the South West Coast of India. Curr. Sci., 40 : 161-162.

Poernomo, A. and Hamami, E. 1983. Induced gonad maturation, spawning and hatching of eye-ablated pond-grown *P. monodon* in a recirculated water environment. Proc. Int. Conf. on Warm Water Aquaculture, Crustacea, 1 : 412 - 419.

Ponnuchamy, R., Reddy, S.R. and Katre, S. 1981. Effects of eyestalk ablation on growth and food conversion efficiency of the freshwater prawn *Macrobrachium lanchesteri* (de Man). Hydrobiologia, 77 : 77 - 80.

Prakash, S. and Agarwal, G.P. 1989. A report on the food and feeding habits of freshwater prawn, *Macrobrachium choprai,* Indian. J. Fish., 36 : (3) : 221 - 226.

Primavera, J.H. 1985. A review of maturation and reproduction in closed-thelycum *penaeids. In:* Proc. of 1st Int. Conf. on the Culture of *Penaeid* Shrimp (1984). SEAFDEC Aqua. Dept.

Prosser, C.L. 1990. Temperature. In : Comparative animal physiology (Ed. C.L. Prosser), 4th Edition, W.B. Saunders company, New York, Chichester, Brisbane, Toronto, Singapore. 109-165.

Quackenbush, L.S. and Herrnkind, W.F. 1981. Regulation of moult and gonadal development in the spiny lobster *Panulirus agrus* (Crustacea : Palinuridae) : effect of eyestalk ablation. Comp. Biochem. Physiol., 69A : 523 - 527.

Quigley,M.A., J.F. Cavaletto and W.S. Gardner. 1989. Lipid composition related to size and maturity of the amphipod, *Pontoporeia hoyi*. J. Great Lakes Res. 15, 601-610.

Raanan, Z., Sagi, A., Wax, Y., karplus, I., Hulata, G. and Kuris, A. 1991. Growth, size rank, and maturation of the freshwater prawn, *Macrobachium rosenbergii* : Analysis of marked prawns in an experimental population.

Rachor, E. 1976. Structure, dynamics and productivity of a population of *Nucula nitidosa* (Bivalvia : Protobranchia). Berichte der Deutschen Wissenschaftlichen kommission fur meersforschung, 24 : 296 - 331.

Radhakrishnan, E.V. and Vijayakumaran, M. 1984a. Effect of eyestalk ablation in spiny lobster *Panulirus homarus* (Linnaeus) 1. Moulting and growth. Indian J. Fish., 31 : 130 - 147.

Radhika, P., Surendranath, P. and Ramana Rao, K.V. 1988. In vitro effect of crustacean eyestalk extracts on nitrogen metabolism of marine crab, *Scylla serrata* (Forskal) Nat. Acad. Sci. Lett., 11 (10) : 325 - 326.

Raghavaiah, K., Ramamurthi, R., Chandrasekharam, V. and Scheer, B.T. 1980. Neuroendocrine control of nitrogen metabolism in the Indian field crab, *Oziotelphusa senex senex* (Fabricius - I). End - Products and elimination. Comp. Biochem. Physiol., 67B : 437 - 445.

Rajyalakshmi,T. 1968. The freshwater prawn M. malcolmsonii for use in pond culture. Sea Expt. J., 6 (3).

Rajyalakshmi, T. 1980. Comparative study of the biology of the freshwater prawn *Macrobrachium malcolmsonii* of Godavari and Hooghly river systems. Proc. Indian Natn. Sci. Acad., 46B (1) : 72 - 89.

Ramamurthi, R., Raghavaiah, K., Sreeramachandramurthy, M., Chandrasekharam, V. and Scheer, B.T. 1981. Neuroendocrine control of nitrogen metabolism in the Indian field crab *Oziotelphusa senex senex* Fabricius - III incorporation of label from Radio-amino acids into tissue organic fractions. J. Reprod, Biol. & Comp. Endocrinol., 1 : 75-79.

Ramamurthy, S. and Manikkaraja, M. 1978. Relation between tail and total length and total carapace length for three commercial species of *penaeid* prawns of India. Indian J. Fish., 25 (1 & 2) : 233 - 234.

Raman, K. 1972. Preliminary observations on the culture of *Macrobrachium malcolmsonii* in freshwater ponds. Symposium on Aquaculture as an Industry, Abstract No. 8 : 4.

Raman, K.V., Shakuntala, K., and Reddy, S.R. 1981. Influence of endogenous factors on the pattern of Ammonia excretion in the prawn *Macrobrachium lanchesteri* (de man). Indian. J. Exp. Biol., 19 (1) : 42 - 45.

Ramana Rao, K.V., Jayaprada, P., Padmaja, D.and Reddy, M.S. 1991. Neuroendocrine control of carbohydrate metabolism in the marine carb, *Scylla Serrata* (Frskal). Poceedings of the National Academic of Sciences (India). 61B (4) : 409 - 914.

Randall, D.J. and Wright, P.A. 1987. Ammonia distribution and excretion in fish. Fish physiol. Biochem., 3 : 107 - 120.

Rangneker, P.V., and Madyastha, M.N. 1969. Effect of cyestalk removal on the rate of oxygen consumption in the crab *Scylla Serrata*. J. Am., Morpho. Physiol., 17 : 84 - 88.

Rangneker, P.V., Sabnis, P.B. and Nirmal, H.B. 1971. The occurrence of a hypoglycemic factor in the eyestalk of the freshwater crab, *Paratelphusa jacguemontil*. J. Anim. Morphol. Physiol., 3 : 137.

Rangneker, P.V., and Madhyastha, M.N. 1969. Effect of eyestalk removal on the rate of oxygen consumption in the crab, *Scylla serrata* (Forskal). J. Anim. Morphol., 16 : 84 - 88.

Rangneker, P.V., and Madhyastha, M.N. 1971. Effect of eyestalk ablation on carbohydrate metabolism of the prawn, *Metapenaeus monoceros* (Fabricus). Indian. J. Exp. Biol., 9 : 462 - 464.

Rangneker, P.V. and Deshmukh, R.D. 1968. Effect of eyestalk removal on the ovarian growth of the marine crab, *Scylla serrata* (Forskal). J. Anim. Morph. Physiol., 15 : 116 - 126.

Rao, K.R., Fingerman, S.W. and Fingerman, M. 1973. Effects of exogenous ecdysones on the moult cycles of fourth and fifth stage American lobster, *Homarus americanus*. Comp. Biochem. Physiol., 44A : 1105 - 1120.

Rao, K.P. 1958. Oxygen consumption as a function of size and salinity in *Metapenaeus monoceros* (Fab.) from the marine and brackishwater environs. J. Exp. Biol., 35 (2) : 307 - 313.

Reddy, P.S. and Kishori, B. 2001. Methionine - Enkephalin induces hyperglycemia through eyestalk hormones in the estuarine crab *Scylla Serrata*. Biol. Bull., 201 : 17-25.

Reddy, P.S. 1990. Neuroendocrine control of metabolism in the freshwater field crab *Oziotelphusa senex* Fabricius. J. Crustac. Biol., 10 (4) : 595-607.

Reddy, D.C. Dratnal, E. and Davies, R.W. 1992. Dynamics of energy storage in a freshwater predator (*Nephelopsis obscura*) following winter stresses. Can. J. Fish. Aquat. Sci., 49 ; 1583-1587.

Reddy, P.S., Devi, M., Sarojini, R., Nagabhushanam, R. Fingerman, M. 1994. Cadmium chloride induced hyperglycemia in the red Swamp Crayfish, procambarus clarkii : Possible role of crustacean hyperglycemic hormone. Comp. Biochem. Physiol., 107 C : 57-61.

Reddy, P.S., Katyayani, R.V. and Fingerman, M. 1996. Cadmium and naphthalene - induced hyperglycemia in the fiddler crab, Uca Pugilator : differential modes of action on the neuroendcrine system. Bull. Environ. Contam. Toxicol., 56 : 425-431.

Reddy, D.C and Davies, R.W. 1993. Metabolic adaptations by leech *Nephelopsis obscura* during longterm anoxia and recovery. The Journal of Experimental Zoology, 265 : 224-230.

Reddy, O.R., Gopal Rao, K., Rama Rao, P.V.A.N. and Ramakrishna, R. 1985. Polyculture of freshwater prawn *Macrobrachium malcolmsonii* (Milne Edwards), major Indian and Chinese Carps. J. of Morphol., 146 : 55 - 80.

Regnault, M. 1979. Ammonia excretion of the sand shrimp *Crangon crangon* (L.) during the moult cycle. J. Comp. Physiol., 133 : 199 - 214.

Regnault, M., 1987. Nitrogen excretion in Marin and Freshwater crustacea. Biol. Rev., 62 : 1 - 24.

Reiss, M.J. 1989. The allometry of Growth and Reproduction of Juvenile normone titers in *Apis mellifera*. Insect Biochem., 17: 1003 - 1007.

Reitman, S. and Frankel, S. 1957. A Colorimetric method for the determination of serum glutamic oxaloacetate and glutamic pyruvate transaminases. Am. J. Clin. Path., 28 : 56 - 63.

Renaud, L. 1949. Le cycle de reserves organiques chez les Crustaces Decapodes. Ann. inst. Oceanog., 24 : 259 - 357.

Rice, J.A. and P.A. Cochran.1984. Independent evaluation of a bioenergetics model for largemouth bass. Ecology. 65 : 732 - 739.

Richardson, A.J., Lamberts, C., Is aacs, G., Moloney,C.L. and Gibbons. M.J. 2000. Length - Weight Relationship of Some Inportant Foruge Crustaceans from South Africa. Naga, The ICLARM Quarterly. 23 (2) : 29 - 33.

Robert, R.M. and Bocyuen. 1974. "Chemical modification of plasma membrane polypeptides of cultural mammalian cell as an aid of studying protein turnover". Biochem., 13 : 4846-4851.

Robertson, L., Thomas, P. and Arnold, C. R. 1988. Plasma cortisol and secondary stress response of cultured red drum *(Sciaenops ocellatus)* to several transportation procedures. Aquaculture, 68:115-130.

Rodriquez, L.M. 1979. A guide to the mass scale production of *Penaeus monodon* spawners. SEAFDEC Aquaculture Dept., Tigbauan, Philippines. 10.

Rosas, C., Cuzon, G., Gaiola, G., Le Priol, Y., Pascual, C., Rossignyol, J., Contreras, F., Sanchez., A. and Van Wormhoudt, A. 2001. Metabolism and growth of juveniles of *Litopenaeus vannamei* : effect of salinity and dietary carbohydrate levels. J. of Expt. Mar. Biol. and Ecol., 259 : 1 - 22.

Rosas, C., Fernandez, I., Brito, R. and Iglesia, E.D. 1993. The effect of eyestalk ablation on the energy balance of the pink shrimp, *Penaeus notialis*. Comp. Biochem. Physiol. 104A (1) : 183 - 187.

Rosas, C., Vanegar, C., Alcaraz, G., and Diaz, F. 1991. Effect of eyestalk ablation on oxygen consumption of callinectes similis exposed to salinity changes. Comp. Biochem. Physiol., 100 A (1) : 75 - 80.

Rotallant, G., Takac, P., Liu, L., Scott, G.L. Laufer, H. 2000. Role of ecdysteroids and methyl farnesoate in morphogenesis and terminal moult in polymorphic males of spider crab *Libinia emarginate*. Aquaculture, 190 : 103 - 118.

Sacyanan, E.B. and Hirata, H. 1986. Circadian rhythm of feeding and respiration in Kuruma prawn, *Penaeus japonicus*. Mini Rev. Data File fish. Res. Lab. Kagoshima Univ., 4 : 63 - 76.

Sadok, S., Uglow, R.F. and Haswell, S.J. 1997. Haemolymph and mantle fluid ammonia and ninhydrin positive substances variations in salinity-challenged mussels (*Mytilus edulis L.*). Journal of Experimental Marine Biology and Ecology, 211 (2) : 195 - 212.

Saenz. F., Garcia, U. and Arechiga, H. 1997. Modulation of electrical activity by 5-Hydroxytryptamine in crayfish neurosacretory cells. The Journal of Experimental Biology, 200 : 3079 - 3090.

Sagi, A., Shoukrun, R., Levy, L., Barki, A., Hulata, G. and Karphes, I. 1997. Reproduction and Molting in previously spawned and firsttime spawned red-claw crayfish *Cherax quadricarinatus* females following eyestalk ablation during the winter reproductive arrest period. Aquaculture, 156 : 101 - 111.

Sampath Kumar, J.S., Nagarathinam, N. and Sunderaraja, V. 2000. Production characteristics of *Macrobrachium malcolmsonii* under controlled mono culture system. J. Aqua. Trop., 15 (3) : 207 - 217.

Santos, E.A., and Keller, R. 1993. Crustacean hyperglycemic hormone (CHH) and the regulation of carbohydrate metabolism : current perspectives. Comp. Biochem. Physiol. 106A : 405 - 411.

Santos, E.A., Nery, L.E.M., Keller. R. and Goncalves, A.A. 1997. Evidence for the involvement of the crustacean hyperglycemic hormone in the regulation of lipid metabolism. Physiol. Zool., 70: 415 - 420.

Sarojini, R., Nagabhushanam, R. and Fingerman, M. 1996. *In vivo* assessment of opioid agonists and antagonists on ovarian maturation in the red Swamp crayfish, *Procambarus clarkii*. Comp. Biochem. Physiol., 115C : 149-153.

Sarojini, R., Mirajkar, M.S. and Nagabhushanam, R. 1982. Annual reproductive cycle and the effect of temperature on the reproduction of the freshwater prawn, *Macrobrachium kistnensis*. Biology, III : 1 - 5.

Sarojini, R., Gyananath, G. and Nagabhushanam, R. 1983. Reproduction and moulting in the freshwater prawn, *Macrobrachium lamarrei*. Comp. Physiol. Eco., 8 : 309 - 312.

Sarojini, R., Nagabhushanam, R. and Fingerman, F. 1997. An *in vitro* study of the inhibitory action of methionine enkephalin on ovarian maturation in the red Swamp crayfish, *Procambarus clarkii*. Comp. Biochem. Physiol., 117C : 207-210.

Sarojini, R., Nagabhushanam, R., and Fingerman, M. 1994. 5-Hydroxytryptaminergic control of testes development through the androgenic gland in the red swamp crayfish, *Procambarus clarkii*. Invert. Reprod. Develop., 26 (2) : 127 - 132.

Savitz, J.1971. Nitrogen excretion and protein consumption of the bluegill sunfish (*Lepomis macrochirus*). J. Fish. Res. Board. Can., 28, 449 - 451.

Savitz. J., Albanese,E., Evinger, M.J. and Kolasinski,P. 1977. Effect of ration level on nitrogen excretion nitrogen retention and efficiency of nitrogen utilization for growth in largemouth bass (*Micropterus salmoides*). J. Fish. Biol., 11 : 185 - 192.

Schatzlein, F.C. and Costlow Jr, J.D., 1978. Oxygen consumption of the larvae of the decapod crustaceans. *Emerita talpoidia* (say) and *Libinia emarginata* Leach. Comp. Biochem. Physiol., 61A : 441-450.

Schmidt - Nielsen, K. 1984. Scaling : Why is Animal Size So Important? Cambridge : Cambridge University Press.

Schoffeniels, E. and Gilles, R. 1970. Nityrogen constituents and Nitrogen Metabolism in Arthropoda. In : Chemical Zoology, Eds, Flarkin. M. and Scheer. Bot. Academic Press, New York, London, 5 : 199 - 227.

Schwoch, G. 1972. Some studies on biosynthesis and function of trehalose in the crayfish, *Orconectes limosus*. Rafinsque. Comp. Biochem. Physiol., 43B : 905 - 917.

Sebens, K.P. 1979. The energetics of asexual reproduction and colony formation in benthic marine invertebrartes. American zoologist, 19: 683 - 697.

Segal, H.L., G.A. Dunway and J.R. Winkler. 1976. The role of lysosomes in protein turnover. In: Proteolysis and Physiological regulation, (Vol. 2, Ed: D.W.Robbins and K. Brew, Academic press, London), 287-308.

Sehnal, F. 1971. Endocrines of arthropods. In : Chemical Zoology, eds. Florkin, M. and Scheer, B.T., Academic Press, New York, London, VI : 307 - 345.

Sheerboom, J.E.M. 1978. The influence of food quantity and food quality on assimilation, body growth and egg production in the pond snail, *Lymnaea stagnalis* (L.) with particular attention to the haemolymph - glucuose concentration. Proceedings of the known like Naederlandse Akadense van Wetes chapper, Series C81 : 184 - 197.

Sheerboom, J.E.M. and Geldof, A.A. 1978. A quantitative study of the assimilation of different diets in the pond snail, *Lymnaea stagnalis* (L), introducing a method to prevent coprophagy. Proceedings of the koninklijke Nederlandse Akademie Van Westenschappen, series C81 : 173 - 183.

Shewbart, K.L., Mies, W.L., and Ludwing, P.D. 1972. Indentification and quantitative analysis of the amino acids present in protein of the brown shrimp, Penaeus azatecus. Mar. Biol., 16 : 64 - 67.

Sibly, R. 1981. Strategies of digestion and defecation. Pages 109-142 in P.calow and C.Townsend, eds. Physiological ecology : an evolutionary approach to resource use. Sinauer, Sunderland, Mass.

Sibly, R.M. and Calow, P. 1986. Physiological Ecology of animals :An evolutionary approach. Block well scientific publications, oxford.

Sierra Uribe, E., Diaz - Herrera, F. and Bueckle - Ramirez, L.F. 1997. Effect of unilateral eyestalk ablation on the physiological energetics of *Procambarus clarkii* (Decapoda, Cambaridae), Riv. Ital. Acquacott., 32 (3) : 105 - 113.

Sierra, E., and Diaz, F. 1999. Dynamics of bioenergetics of post larval and juveniles of *Macrobrachium rosenbergii* caused by unilateral eyestalk ablation. Journal of Aquaculture in the tropics. 14 (2) : 113 - 119.

Silas, E.G. 1982. Major breakthrough in spiny lobster culture. Mar. Fish. Infor. Serv. T and E . Ser., 43 : 1 - 5.

Sindhus Kumari, S. and Pandian, T.J. 1987. Effect of unilateral eyestalk ablation on moulting, growth, reproduction and energy budget of *Macrobrachium nobilii*. Asian Fish. Sci., 1 : 1 - 17.

Skinner, D.M. 1965. Amino acid incorporation into protein during the moult cycle of the land crab, Gercarcinus lateralis. J. Exp. Zool., 160 : 226-233.

Skinner, D.M. 1984. Moulting and regeneration. In : The Biology of Crustacea, (eds. Bliss, D.E. and L.H. Mantel), Academic Press, New York. 9 : 43 - 146.

Smith, R.I. 1940. Studies on the effect of eyestalk removal upon young crayfish (*Cambarus clarkii* Girard). Biol. Bull., 79 : 145 - 152.

Smith, K.L., Jr. 1973. Energy transformation by the sargassum fish, *Histrio histrio* (L). Journal of experimental marine biology and ecology. 12 : 219 - 227.

Smith, M.A.K. 1981. Estimation of growth potential by measurement of tissue protein synthetic rates in feeding and fasting rainbow trout, *Salmo gairdneri* Richardson. Journal of fish Biology. 19 : 213 - 220.

Smith, R.R., Rumsey, G.L. and Scott, M.L. 1978. Heat increment associated with dietary protein, fat, carbohydrate and complete diets in salmonids: comparative energetic efficiency. J. Nutr., 108 : 1025 - 1032.

Smith, D.E.C. and Davies, R.W. 1995. Effects of feeding frequencies on energy partitioning and life history of the leech *Nephalopsis obscura*. J.N. Am. Benthol. Soc., 14 (4) : 563 - 576.

Sochasky JB, Aiken, D.E. and McLeese, D.W. 1973. Does eyestalk ablation accelerate molting in the lobster *Homarus americanus*? J. Fish. Res. Board Can., 30 : 1600 - 1603.

Solorzano, L. 1969. Determination of ammonia in natural waters by the phenol-hypochlorite method. Limnol. Oceanogr., 14 : 799 - 801.

Soumoff, C. and O' Connor, J.D., 1982. Repression of Y-organ secretory activity by moltinhibiting hormone in the crab, *Pachygrapsus crassipes*, Gen. Comp. Endocrinol., 48 : 432 - 439.

Soyez, D. and Kleinholz, L.H. 1977. Moult inhibiting factor from the crustacean eyestalk. Gen. Comp. Endorcrin., 31 : 233 - 242.

Spaziani, E., Rees, H.H., Wang, W.L. and Watson, R.D. 1989. Evidences that the Y-organ of the crab *Cancer antennarius* secrete 3-hydro ecdysone. Mol. cell. Endocrinol., 66 : 17 - 25.

Sreedhar, S.K and C.K. Radhakrishnan. 1995. Energy storage and utilization in relation to gametogenesis in the mussel, *Musculista senhausia* (Bivalvia : Mytilidae) from cochin backwaters, West coast of India. Indian Journal of marine sciences. Vol. 24, pp. 206.

Srikantan, T.N. and Krishna murthy, C.R. 1955. Tretrazolium test for dehydrogenases. J. Sci, Indust. Res., 14: 206.

Steele, J.H. 1974. The structure of Marine ecosystems. Cambridge, Mass., Harvard University Press.

Stephenson, M.J. and Knight, A.W. 1980. The effect of temperature and salinity on oxygen consumption of post-larvae of *Macrobrachium rosenbergii* (De Man) (Crustacea : Palaemonidae). Comp.Biochem. Physiol., 67 : 699 - 703.

Stevenson, J. 1972. Changes in activities of the crustacean epidermis during the molting cycle. Am. Zoologist., 12 : 373 - 380.

Stewart, D.J., Weiniger, D., Rottiers, D.V. and Edsall, T.A. 1983. An energetics model for lake trout, *Salvelinus namaycush* : application to the Lake Michigan population. Canadian Journal of Fisheries and Aquatic Sciences. 40 : 681 - 698.

Stewart, D.J. and Binkowski, F.P. 1986. Dynamics of consumption and food conversion by lake Michigan alewifes: an energetics-Modelling synthesis. Transactions of the American Fisheries Society. 115 : 643 - 661.

Studier, E.M., Edwards,K.E. and Thompson, M.D. 1975. Bionenergetics in two pulmonate snails, *Helisoma* and *physa*. Comp. Biochem. Physiol., 51A : 859 - 861.

Suarez, A.G. and Xiques, R.D. 1969. Consideraciones sobre los indices metabolicos y la supervivencia del camaron blanco penaeus schmitti, Burkenroad, de la plataforma Cubana. FAO Fish. Rep., 57 (3) : 621 - 642.

Subrahmanyam, M. 1973. Fishing and biology of *M.monoceros* (Fabricius) from Godavari estuarine system. Bull. Brackish water aquaculture in India, 31.

Surendranath, P., Ramanaiah, K. and Rao, K.V.R. 1992. Effect of eyestalk extract and kelthane on penacid prawan *Metapenaeus monoceros* (Fabricius). Indian Journal of Eperimental Biology, 30 (8) : 676 - 679.

Surendranath, P., Kasaiah, A., Sekhar. V., Surendra Babu, K. and Ramana Rao, K.V. 1987. Starvation induced alterations in oxygen consumption and organic constituents of penaeid prawn, *Penaeus indicus* (H.M. Edwards) Maha Sagar, 20 (3) : 199 - 203.

Swami, K.S., K. S. Jagannatha Rao, K. Satyavelu reddy, K. Sreenivasa Moorthy, K. Lingamurthy, C.S. Chetty and K. Indira. 1983. The possible metabolic diversions adopted by the fresh water mussel to counter the toxic metabolic effects of selected pesticide: Ind.J.Comp. Anim. Physiol., 1:95- 101.

Swaminathan, M. 1983. Hand book of food and nutrition. 3rd edition : pp. 22.

Tagatz, M.E. 1968. Growth of juvenile blue crabs, *Callinectes Sapisus* Rathbun, In the St. Johns River, Florida. Fishery Bull. Fish. Wildl. Serv. US., 67 : 281 - 288.

Taketomi, Y. and Kawano, Y. 1985. Ultrastructure of the mandibular organ of the shrimp, *Penaeus japonicus,* in untreated and experimentally manipulated individuals. Cell. biol. Int. Rep., 9 : 1069 - 1074.

Taketoni, T. Motono, M. and Miyawaki, M. 1989. On the function opf the mandibular gland of decapod crustacea. Cell Biol. Internat. Rep. 13 : 463 - 469.

Tan-Fermin, J.D. 1991. Effects of unilateral eyestalk ablation on ovarian histology and oocyte size frequency of wild and pond-reared *Penaeus monodon* (Fabricius) brood stock. Aquaculture, 93 : 77 - 86.

Tandler, A. and Beamish, F.W.H. 1981. Apparent specific dynamic action (SDA), fish weight and level of caloric intake in largemouth bass, *Micropterus salmoides* Lacepede. Aquaculture, 23 : 231 - 242.

Tandler, A. and Beamish, F.W.H. 1980. Specific dynamic action and diet in largemouth bass, *Micropterus salmoides* (Lacepede). Journal of Nutrition. 110 : 750 - 764.

Tandler, A. and Beamish, F.W.H. 1979. Mechanical and biochemical components of apparent specific dynamic action in largemouth bass, *Micropterus salmoides* Lacepede. Journal of fish biology. 14 : 343 - 350.

Tavill, A.S. and Cooksley, W.G.S. 1983. Biochemical aspects of liver disease. In : Biochemical aspects of human disease, (eds) R.S. Elkeles, A.S. Tavill, Black-well Scientific publications, Boston.

Tayyabha, K., M. Hasan, I. Fakhrul and N.H. Khan. 1981. Organophosphate pesticide metasystox induced regional alterations in brain nucleic and metabolism. Ind. J. Exp. Biol., 19: 688-690.

Telford, M. 1968. Changes in blood sugar composition during the molt cycle of the lobster, *Homarus americaurus*. Comp. Biochem. Physiol., 26, 917 - 926.

Thomas, M.M. 1974. Reproduction, fecundity and sex ratio of the green tiger prawn, *Penaeus semisulcatus* (De Haan). Indian J. Fish., 2 : 152 - 163.

Thompson, J.M., E.P. Bergersen, C.A. Carlson and L.R. kaeding. 1991. Role of size condition and lipid content in the overwinter survival of Age-0 Colorado squawfish. Transactions of the American fisheries Society. 120: 346-353.

Thornborough, J.R. 1968. Neuroendocrine repression of ribonuclease in the prawn, *Palaemonetes Vulgaris*. Comp. Biochem. Physiol., 24 : 625 - 628.

Ting, Y.Y. 1970. Study on the oxygen consumption of grass shrimp, *Penaeus monodon* and shrimp, *Metapenaeus monoceros*. Bull. Taiwan. Fish. Res. Inst., 16 : 111 - 118.

Toneys, M.L. and D.W. Coble. 1979. Size related, first winter mortality of freshwater fishes. Transactions of the american Fisheries Society. 108:415-419.

Townsend, C.R. and Calow, P. 1981. Physiological ecology: an evolutionary approach to resource use. Blackwell scientific publications, Oxford.

Travis, D.F. 1957. The moulting cycle of the spiny lobster, *Panulirus argus* Latreille. IV. Post-ecdysial histological and histochemical changes in the hepatopancreas and integumental tissues. Biol. Bull., 113 : 451-479.

Travis, D.F. 1954. The moulting cycle of the spiny lobster, *Panulirus argus* Latreille. I. Moulting and growth in laboratory maintained individuals. Biol. Bull., 107 : 433 - 450.

Treece, G.D. and Fox, J.M. 1993. Design, Operation and Training Manual for an Intensive Culture Shrimp Hatchery, with Emphasis on *P. monodon and P. vannamei*. Texas A&M Univ., Sea Grant College Program, Bryan, Texas, Pub. 93 - 505 - 187 p.

Truchot, J.P. 1983. Regulation of acid - base balance. In the biology of crustacea, Vol.5. Internal Anatomy and physiological regulation (ed. L.H. Mental), pp.431 - 457, New York : Academic press.

Stevenson, J.R. 1968. Metecdysis molt staging and changes in the cuticle in the crayfish Orconectes sanborni (Faxon). Crustaceana., (Leiden) 14 : 169 - 177.

Tsukimura, B. and Borst, D. W. 1992. Regulation of methylfarnesoate in the hemolymph and mandibular organ of the lobster, *Homarus americanus*. Gen. Comp. Endocrinol., 86 (2) : 297 - 303.

Urich, K. 1967. Amino Saurestoff wechsel in des flusskrebss oroconectes limosus. Transaminerung, oxidative and nicht oxydative Desaminierung. Z. Vergl. Physiol., 56 : 95-110.

Vaca, A.A., and Alfaro, J. 2000. Ovarian maturation and Spawning in the white shrimp, *Penaeus vennamei*, by serotonin injection, Aquaculture, 182 : 373 - 385.

Vahl, O. and Davenport, J. 1979. Apparent specify dynamic action of food in the fish *Blennius Pholis*. Marine ecology progress sries. 1 : 109 - 113.

Van Herp, F. 1998. Molecular, Cytological and Physiological aspects of the crustacean hyper glycemic hormone family. In : Coast, G.M., Webster, S.G. (Eds.), Recent advances in Arthropod Endocrinology. Cambridge Univ. Press, Cambridge, pp. 53 - 70.

Vernberg, F.J. 1983. Respiratory adaptations in "The biology of crustacea" (vernberg, F.J. and Vernberg, W.B. eds.). Vol.8 : 1 - 42. Academic Press, New York.

Vernberg, F.J. 1983. Respiratory adaptations in "The biology of crustacea" (vernberg, F.J. and Vernberg, W.B. eds.). Vol.8 : 1 - 42. Academic Press, New York.

Vernberg, W.B., and Vernberg, F.J. 1972. Environmental Physiology of Mrine animals. Springer verlag. Berlin and New York.

Vernberg, W.B., Moreira, G.S. and Mc Namara, J.C. 1981. The effect of temperature on the respiratory metabolism of the developmental stages of *Pagurus criniticornis* (Dana) (Anomura : paguridae). Mar. Biol. Lett., 2 : 1 - 9.

Vijayakumaran, M. and Radhakrishnan, E.V. 1984. Effect of eyestalk ablation in the spiny lobster *Panulirus homarus* (Linnaeus). 2. On food intake and conversion. Indian. J. Fish., 31 (1) : 148 - 155.

Vijayan, K.K., Sunilkumar Mohamed, K. and Diwan, A.D. 1997. Studies on moult staging, moulting duration and moulting behaviour in Indian white shrimp *Penaeus indicus* Milne Edwards (Decapoda : Penaedae). J. Aqua. Trop., 12 (1) : 53 - 64.

Von Bertalanffy, L. 1938. A quantitative theory of organic growth. Human Biology, 10 : 181 - 213.

Vosloo, A., Van Aardt, W.J. and Mienie, L.J. 1996. Presence of itaconic acid in the hemolymph and tissues of the freshwater crab, *Potamonautes warreni* Calman. Comparative Biochemistry, 113B(4):823-825.

Wainwright, G., Webster, S.G. Wilkinson, M.C., Chung, J.S. and Rees, H.H. 1996. Structure and significance of mandibular organ inhibiting hormone in the crab, *Cancer pagurus*. J. Biol. Chem., 271 : 12740 -12754.

Wajsbort, N., Krom, M.D., Gasith, A. and T. Samocha. 1989. Ammonia excretion of green tiger prawn, *Penaeus semisulcatus* as a possible limit on the biomass density in shrimp ponds. Bamidgeh, 41 : 159 - 164.

Wareen, C.E. and Davis, G.L. 1967. Laboratory studies on the feeding of fishes. In: The biological basis of freshwater fish production (ed.S.D. Gerking), pp.175-214. Blackwell scientific publications, Oxford.

Webster, S.G. 1985. The effect of eyestalk removal wounding and limb loss upon moulting and proecdysis in the prawn *Palaemon elegans* (Rathke). J. Mar. Biol. Assoc. U.K., 65 : 279 - 292.

Webster, S.G. 1998. Neuropeptides inhibiting growth and reproduction in crustaceans. In : Coast, G.M., Webster, S.G. (Eds), Recent Advances in Arthropod Endocrinology, Cambridge University Press, 33 - 52.

Wedemeyer, G. A. 1972. Some physiological consequences of handling stress in the juvenile coho salmon *(Oncorhynchus kisutch)* and steelhead trout *(Salmo gairdneri)*. Journal of the Fisheries Research Board of Canada, 29:1780-1783.

Wenner, A.M. 1972. Sex ration as a function of size in marine crustacea. Amer. Natural., 196 : 321 - 350.

Wenzel, F., Meyer, E. and Schwoerbel, J. 1990. Morphometry and Biomass determination of dominant mayfly larve (Ephemeroptera) in running waters. Arch. Hydrobiol., 118 : 31 - 46.

Whiteledge, G.W., Hayward, R.S., Noltie, D.B., Ning-Wang. and N-Wang. 1998. Testing bioenergetics models under feeding regimes that elicit compensatory growth. Transactions of the American Fisheries Society. 127 (5) : 740 - 746; 32 ref.

Wickins, J.F. 1985. Ammonia production and oxidation during the culture of marine prawns and lobsters in laboratory recirculation systems. Aquacult. Engin., 4 : 155 - 174.

Wieser, W. & Medgyesy, N. 1990. Aerobic maximum for growth in the larvae and juveniles of a cyprinid fish, *Rutilus rutilus* (L) : Implications for energy budgetting in small poikilotherms, Functional Ecology, 4 : 233 - 242.

Wieser, W. 1994 : Cost of growth in cells and organisms. General rules and comparative aspects. Biological Reviews, 68 : 1 - 33.

Wightman, J.A. 1981. Why insect energy budgets do not balance. Oecologia (Berlin) 50 : 116-169.

Wilder, M.N., Okumura, T., Suzuki, Y., Fusetani, N. and Aida, K. 1994. Vitellogenin production induced by eyestalk ablation in Juvenile Giant Freshwater prawn Macrobrachium rosenbergii and trial methyl farnesoate administration. Zool. Sci., 11 : 45 - 53.

Williams, M.J., Veith, L.C. and Correll, R.L. 1980. Models for describing shape and allometry, illustrated by studies of Australian Species of *Uca* (Brachyura, Ocypodidae). Australian Journ. Mar. Freshwater Res., 31 : 757 - 781.

Wilson, J.G. 1988. Resource partitioning and predation as a limit to size in *Nucula turgida* (Leckenby and Marshall). Functional Ecology, 2 : 63 - 66.

Windell, J.T. 1978. Digestion and the daily ration of fishes. In: Gerking, S.D. (ed.). Ecology of freshwater fish production. pp. 159-183. Oxford, Blackwell.

Yamamoto, K.I. 1991. Relationship of respiration to body weight in the carp, *Cyprinus carpio* under resting and normoxic condition.

Yamaoka, L.H. and Scheer, B.T. 1970. Chemistry of growth and development in crustaceans. Chemical Zoology, Vol.5, Academic Press, New York, 321 - 341.

Yudin, A.I., Diner, R. A., Clark, W.H. and Chang, E.S. 1980. Mandibular gland of the blue crab *Callinectes sapidus*. Biol. Bull. 159. 760 - 772.

Zandee, D.I. 1966. Metabolism in the crayfish, *Astacus astacus* (L) . II The energy yielding metabolism. Arch. Intern. Physiol. Biochem., 74 : 45 - 47.

Zar, J.H. 1984. Biosatistical analysis. 2nd end. Pentice-Hall, Engle wood cliffs, NJ :

Zebe, E., Roters, F.J. and Barbara, K. 1986. Metabolic changes in the medical leech, *Hirudo medicinalis*, following feeding. Comp. Biochem. Physiol., 84A : 49 - 55.

Zeleny, C. 1905. Compensatary regulation. J. Exp. Zool., 2 : 1 - 102.

Zeuthen, E. 1970. Rate of living as related to body size in organisms. Pol. Arch. Hydrobiol., 17 : 21 - 30.

Zeuthen, E. 1953. Oxygen uptake as related to body size in organisms. Q. Rev. Biol., 28 : 1 - 12.

www.ingramcontent.com/pod-product-compliance
Lightning Source LLC
Chambersburg PA
CBHW032006170526
45157CB00002B/566